LARGE PRINT

DATE DUE

AuG 1, 2018	
DISCARDED	
	PRINTED IN U.S.A.

TRAVELS WITH CASEY

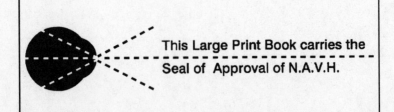

This Large Print Book carries the
Seal of Approval of N.A.V.H.

TRAVELS WITH CASEY

BENOIT DENIZET-LEWIS

THORNDIKE PRESS
A part of Gale, Cengage Learning

GALE
CENGAGE Learning·

Farmington Hills, Mich • San Francisco • New York • Waterville, Maine
Meriden, Conn • Mason, Ohio • Chicago

GALE
CENGAGE Learning®

LIBRARY OF CONGRESS CATALOGING-IN-PUBLICATION DATA

Denizet-Lewis, Benoit, author.
   Travels with Casey / by Benoit Denizet-Lewis. — Large print edition.
     pages cm — (Thorndike Press large print nonfiction)
     Originally published: New York : Simon & Schuster, 2014.
     Includes bibliographical references.
     ISBN 978-1-4104-7151-2 (hardcover) — ISBN 1-4104-7151-9 (hardcover)
     1. Dogs—United States. 2. Dogs—Effect of human beings on—United States. I. Title.
  SF422.6.U6D46 2014b
  636.7—dc23   2014023231

Published in 2014 by arrangement with Simon & Schuster, Inc.

Printed in the United States of America
1 2 3 4 5 6 7 18 17 16 15 14

To cats.
(Just kidding.)
To dogs!

Travels with Casey

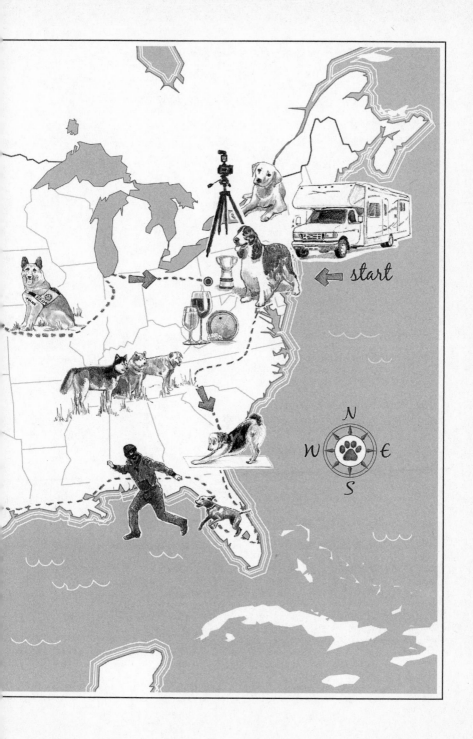

start

CONTENTS

AUTHOR'S NOTE

In most cases, I have included the full names of those I met on my journey. In some instances, I used only a first name or changed a person's name.

For narrative purposes, I occasionally altered the sequence of dogs or people I met in a particular city. For space reasons, I also had to leave out many amazing dogs and humans I encountered during my travels.

PROLOGUE

"I don't think my dog likes me very much,"
I told Dr. Joel Gavriele-Gold, a Manhattan
psychoanalyst shaped like an English Bulldog.

13

I had come to interview Dr. Gold about his book, *When Pets Come Between Partners: How to Keep Love — and Romance — in the Human/Animal Kingdom of Your Home,* but he had deftly turned the tables and was now mining my subconscious for signs of canine-related dysfunction. Against my better judgment, I confessed to worrying that my nine-year-old Lab/Golden Retriever mix, Casey, would prefer to live elsewhere. By elsewhere, I didn't mean in a different house or in a different neighborhood. I meant with a different human.

"I feel like I'm disappointing Casey, like I'm not good enough for him," I explained to Dr. Gold, who specializes in the myriad ways we saddle our pets with our psychological baggage.

He leaned forward in his black office chair and looked at me with what felt like grave concern. "Go on," he said, his gentle gaze moving down to Casey, who lounged at my feet on the carpeted floor of this garden-level office decorated with old paintings of dogs (including one of Rin Tin Tin). A few feet away, Dr. Gold's enormous black herding dog, Dova, snored on a sofa by the window. Dr. Gold believes that dogs make for valuable therapeutic tools, and in the last fifteen years he has rarely come to work

14

without one.

But I wasn't sure I wanted to go on. I was embarrassed to be here, prattling on about how my dog hurts my feelings — about how he makes me question whether I'm a good person. Dogs, after all, have been carefully bred to make us feel like champions. Best of all, we don't have to do all that much to earn their adoration. This is a critical difference between dogs and cats. Mistreat your cat, and he will likely shit on your pillow and find a better place to live. Mistreat your dog, and he will likely stick around — eager to let bygones be bygones. Dogs are relentless optimists, eager to lick your face at the slightest sign of your better self.

Of course, I don't mistreat Casey. We swim in the ocean, chase tennis balls at the park, play hide-and-seek around the house. We wrestle. We bark at the moon. We even have our special little game, where I ask him to "Little Speak" and he growls softly and adorably.

Still, I've spent much of our nine years together convinced that he's not especially fond of me. And I didn't think it was *all* in my head. "Casey's really good at looking miserable," I told Dr. Gold.

If my dog deems it too long since his last walk or feeding, he will sigh loudly. (I've never heard a dog sigh the way Casey does. It's his most human trait.) Many dogs seem to play the "poor me" card when they want to guilt us into something, but with Casey the look seems cruelly genuine.

Even when Casey looks happy, he's not one for outward displays of loyalty. He's not like the dog I grew up with, Milou, a Husky/German Shepherd mix named after the dog in the popular Belgian comic *Les Aventures de Tintin.* Milou liked sleeping with her head in my lap and always seemed to be checking up on the humans around her. "Are you okay? Anything I can do?" seemed to be her default doggie mode.

Casey, on the other hand, seems to like my friends more than me. I relayed a heartbreaking story to Dr. Gold: When Casey was two, I'd had to leave my Boston apartment for a week-long business trip. My friend Mike wasn't working at the time and agreed to take temporary custody of my dog. He promised to "spoil him rotten," and he didn't disappoint. I received regular, hyperbolic updates recounting marathon-length walks on the beach, extravagant hamburger doggie dinners, a pet store chew toy spending spree, and an afternoon devoted entirely to letting Casey hump dogs at the dog park.

Upon my return, I went to Mike's apartment to retrieve my "best friend." Casey seemed happy enough to see me, but as I tried to coax him into the car, he ran back to the front door of Mike's apartment building, planted his butt on the stoop, and stared expectantly at the doorknob.

"It's difficult to overstate how crushed I was by this . . . betrayal," I stammered in Dr. Gold's office.

"You saw it as a betrayal?"

"I did."

"It sounds to me like he was just being a Lab," Dr. Gold said, looking at Casey with a warm, rosy-cheeked smile.

"How do you figure?"

"He probably just wanted more hamburgers."

"I don't think this was about hamburgers," I countered, more dramatically than I intended.

"What do you think it was about?"

That was easy. Casey wanted to live with Mike. And was I being selfish by not letting him go? Was I putting my neediness ahead of my dog's happiness?

Those questions had ricocheted around in my head as I called for Casey to come to me from Mike's doorstep. But the animal I'd loved and cared for since he was eleven weeks old looked right through me and didn't move. (Woodrow Wilson once said, "If a dog will not come to you after having looked you in the face, you should go home and examine your conscience.")

Devastated, I marched over to Mike's doorstep, grabbed Casey by the collar, and pulled him toward the car. He looked sad. I must have looked sadder.

In Caroline Knapp's book *Pack of Two,* she writes that "in all likelihood dogs do not make comparative assessments about their lives . . . do not lie around wishing they were elsewhere, fantasizing about better owners, dreaming of more varied settings."

I'm not so sure. I imagine that if Casey were blessed with the power of human speech, he might break my heart. "You've been *interesting,*" I could see him saying, "but I'd like to give another human a try."

Dr. Gold listened patiently as I unburdened myself in his spacious office on Manhattan's Upper West Side. Every now and then he would look down at Casey with fondness, and when he did this I would wonder — as I often do — what my dog was thinking. Did Casey know how much he affected me? Did he realize we were talking about him?

"Do you think I'm crazy?" I asked Dr.

Gold midway through our meeting. "I'd like you to tell me if I'm crazy, because I feel like a crazy person for not just appreciating Casey for the dog he is."

"You're not crazy," Dr. Gold assured me. "It just sounds to me like Casey might not be the dog of your dreams."

Not the dog of my dreams.

I didn't know what to say to that. "I feel guilty for even *thinking* that," I told him.

"You shouldn't," he replied. "Don't think there's any difference between you and the mother of a child. Many mothers are mortified to admit their disappointment that their child isn't the child of their dreams — different temperament, different interests, tuned in differently. Same thing with a dog."

Dr. Gold shifted his attention to Dova, a Bouvier des Flandres (a Belgian herding dog known for its big head, beard, and mustache). "She's not the dog of my dreams, either," he said. "I mean, she's a perfectly nice dog. I care for her and care about her, but she doesn't have that typical *I can't get enough of you* Bouvier vibe. And it sounds like Casey doesn't have that same typical Lab vibe."

I nodded. "I wish Casey was more of a cuddler."

"I wish Dova was, too," Dr. Gold agreed.

19

"She sleeps on the floor at night, and about ten minutes before my alarm goes off, she sneaks quietly on the bed and puts her head on my shoulder. She pretends like she's been there all night!"

He let out a hearty laugh. Here was a man who didn't take his dog's quirks personally.

"But all you hear from pretty much everyone is how much unconditional love they get from their dog," I said. "I don't feel it."

Dr. Gold shook his head. "Not everyone. People have a million kinds of conflicts with their dogs. Why do you think so many dogs end up in shelters? We talk about how great dogs have it in this country, sleeping in our beds and being members of our families. But we kick dogs to the curb, too. We treat them poorly or give them up if they're too much trouble. And it's rarely the dog's fault. People project what they don't like about themselves onto their pet. Or the pet is a kind of psychological stand-in for someone in their lives, or from their past. We think we're mad at the dog, but we're actually mad at our husband, or our dad, or our kid."

"I wonder who I'm secretly mad at," I said.

"I wonder that, too," Dr. Gold replied.

I scratched my nose. "I was *joking.*"

"I know you were," he said. "But if you're

willing, I'd like you to tell me more about your mother."

Earlier in our session, I'd volunteered that I'd spent many years in therapy talking about my mother. Though we have a wonderful and loving relationship today, when I was young, she struggled with depression and could be cold and insensitive. I don't remember her ever hugging me. As I struggled through my twenties, I was pretty sure she'd ruined my life.

Still, I didn't see what my mom had to do with my dog. "I think you might be barking up the wrong tree," I said, pleased with my cleverness.

But he wouldn't let me deflect without a fight. "Growing up," Dr. Gold wondered, "did you often feel like you were disappointing your mother?"

"All the time," I conceded. "I couldn't do anything right. I wanted her to be a normal mother who nurtured me, who made me feel okay about myself."

The room got quiet. "You deserved a mother who would do that," he said a few seconds later, his words heavy with emotion.

"I think I did, too," I mumbled, suddenly on the verge of tears.

"It sounds to me like you didn't get the

love you needed from your mother," he told me, "just like you aren't getting the love you think you need from Casey."

I felt tightness in my chest. Casey yawned.

"What if I were to say to you that Casey is not your mother?" Dr. Gold continued.

"Well, I know that," I stammered.

"Well, you know it *intellectually.* But what is your gut telling you?"

I breathed deeply, leaned back in my chair, and tried to access the wisdom stored in my gut. As I did, I noticed a Sigmund Freud action figure leaning against a book on Dr. Gold's massive wooden bookshelf.

I knew that Dr. Gold had long been obsessed with Freud. In an article he wrote trumpeting the therapeutic value of dogs, he'd recounted a 1971 trip to London, during which he showed up uninvited at the home of Freud's daughter, Anna, who was also a prominent psychoanalyst. He'd brought along one of Anna's books and was hoping to get her autograph, but the elderly housekeeper who answered the door — Sigmund Freud's longtime maid, Paula Fichtl, who had accompanied the Freuds as they fled Vienna and the Nazis in 1938 — said that Anna was on vacation.

Paula invited Dr. Gold inside and led him

to Sigmund Freud's study, where she told him that Freud loved dogs and that they were often at his feet during therapy sessions. She added that Freud once remarked that his Chow Chow, Jo-Fi, understood what a patient needed better than he did.

Dr. Gold never forgot that, and when he rescued a shivering stray Poodle mix from underneath a car in 1973, it wasn't long before the dog — whom he named Humphrey — was accompanying the doctor to work. Humphrey made a "wonderful co-therapist," Dr. Gold recounted in an article for the *AKC Gazette,* the magazine of the American Kennel Club. The dog would lie by the couch next to patients and seemed to instinctively know when they were moving into difficult emotional territory. At those times he would lift his head and place it on a patient's arm, usually prompting the person to "move into ever deeper emotions."

Amos, a Lab and retired seeing-eye dog, came soon after Humphrey and was especially good with kids. When a child wouldn't open up in therapy, Dr. Gold would suggest that the child speak instead to the dog. "I would never make you sit on the radiator with no clothes on," one boy, who had been abused, confided to Amos.

Before long, Dr. Gold began seeing individuals and couples in therapy he believed were "transferring" their feelings and relationship problems onto their pets. In his book, Dr. Gold explores several forms of transference, including "displacement," which he defines as "the unconscious act of putting past thoughts, moods, feelings, images, impulses, and even actions onto present-day individuals." Gold argues that "someone might fear a very large dog because of an association with an overbearing parent."

The longer I spent in Dr. Gold's office, the more I suspected I was using displacement (and probably other psychological coping mechanisms, too) in my relationship with Casey. Was I expecting my dog to make up for my mother's shortcomings, to give me the love she hadn't? When Casey didn't deliver, was I growing judgmental and impatient with him just like my mom had been with me? And when I treated my dog that way, was I feeling the shame that Dr. Gold insisted my mother should have felt about her treatment of me?

Though I'd first rejected Dr. Gold's dog-mother-confusion hypothesis, I now suspected he might be right. Maybe confusing Casey and my mother wasn't that big a stretch; after all, our parents and our dogs might be the only living things from whom we have the right to expect "unconditional love." One truth, though, was beyond doubt: I had some serious work to do on my relationship with Casey.

"I suppose I'll have plenty of time on the road these next few months," I told Dr. Gold.

"Ah, yes, the *road trip,*" he said. "Do you know how fortunate you are to be able to drop everything and drive around the country with your dog?"

Dr. Gold wasn't the first person to express envy at my impending cross-country adventure, during which I'd be traveling for nearly four months through thirty-two states in a rented motorhome with Casey.

My idea — man forces dog into RV, then writes book about it — certainly wasn't original. Practically everyone I'd told about my journey brought up *Travels with Charley,* John Steinbeck's 1962 travelogue about his cross-country voyage with his Poodle. One longtime acquaintance, who says things out loud that normal people merely think privately, cornered me at a party and announced, "Benoit, your book's already been done by a better writer than you."

Maybe so, but the purposes of our trips were different. Steinbeck had traveled the country to tell the story of America's soul; I was traveling the country to tell the story of America's dogs. Charley was an accessory; Casey would have a starring role.

When friends would ask why I wanted to write a book about dogs, I would often say, "Because, of course, there aren't enough books about dogs." In truth, there are probably too few books about cats and too many books about dogs. My bookshelf overflows with tales of troublesome but lovable puppies, inspirational strays, selfless service dogs, and brave canine warriors. Then there are the endless how-to manuals — how to feed your dog, train your dog, walk your dog, groom your dog, and, in the burgeoning age of canine cognition, understand what your dog is thinking, feeling, and dreaming.

I wanted to write a different kind of dog book, one that would explore and celebrate the breadth of human-dog relationships in contemporary life. To do that, I planned to travel across America — the country with the highest rate of dog ownership on the planet — and hang out with as many dogs (and dog-obsessed humans) as I could.

On my journey I planned to meet therapy

dogs, police dogs, shelter dogs, celebrity dogs, farm dogs, racing dogs, stray dogs, show dogs, hunting dogs, dock-diving dogs, and dogs with no discernable "job" other than lounging around the house and terrorizing the mailman.

I would also be hanging out with dog rescuers and shelter workers, canine cognition experts, show dog owners, wolf lovers, dog walkers, dog healers, dog haters, pet photographers, and dog mascot owners. And that was only on the East Coast. On my journey I'd also be cavorting with pet psychics, with my former middle-school English teacher who now works part-time as a "dog masseur," and with a woman who was taken to court for not picking up her dog's poop.

We can't agree on much in our increasingly polarized country, but we tend to find common ground when it comes to our love for canines. What, I wondered, could be learned about modern American life by exploring our relationship to our dogs? How do the ways we treat — or mistreat — our pets help us understand our values?

It doesn't take an advanced psychology degree, though, to realize the most compelling reason I was dropping everything to travel around the country with an animal: I hoped to find some answers about my fraught relationship with Casey. I wanted — I *needed* — to feel better about my dog. Maybe other dog lovers could show me the way.

"I'm going to think of myself as a failure if I get back from this and nothing has changed between me and him," I told Dr. Gold. "This trip is a chance for Casey and I to bond, to get closer, to better understand each other."

"Do you really expect some profound change from this voyage?" he asked.

His skepticism surprised me. Everyone that I'd told about the trip agreed it would be a great bonding experience for Casey and me. Was that wishful thinking?

Dr. Gold looked down at Casey and then back at me. "You and Casey have had eight years to figure each other out," he said. "I just don't want to see you set yourself up for failure if there isn't any seismic shift in your relationship. It may just be another thing for you to feel shame about."

I sighed, prompting Casey to look toward me from his position on the floor. This seemed to animate Dr. Gold.

"See what happened right there?" he asked.

"What do you mean?"

"You sighed, and Casey looked at you and checked you out. I think you downplay, or simply don't notice, all the ways that you guys are already connected. He is deeply connected to you."

I shook my head. "I often sigh right before I'm going to get up," I explained. "If Casey hears a sigh, he thinks I'm about to get up, and if I get up, he might be getting a walk. I don't think he looked at me because he's particularly worried about me."

Dr. Gold laughed in exasperation. "You're as hard on your dog as you are on yourself!" he said. "I'm interested in what's going on in you that you can't accept that you have a real connection to this dog. It seems like you're desperate for Casey to speak and say, 'Hey, Benoit, you're really okay.' "

At that moment, footsteps outside the office prompted Casey to bark loudly, which he's prone to doing when he hears an unexpected noise outside whatever room he's occupying. The disruption even riled up Dova, who Dr. Gold had told me earlier was "mostly deaf." Apparently not. For about ten seconds, our dogs engaged in a barking showdown that must have startled Dr. Gold's next patient. At this, we shared a good laugh.

"Well, our time is almost up," Dr. Gold

said when quiet returned to the room. He stood up, walked over to Casey, and patted him vigorously on the head. "There are many people who would die to do what you're about to do," he told me. "You're a lucky man."

He was right — I was a lucky man. The timing also seemed right. I was recently single. I was in my mid-thirties and child-less. And I didn't know how many more years Casey and I would have together. Though he's often mistaken for a dog half his age, Casey had recently turned nine. In another year or two, a cross-country trip with him might not be possible.

Still, as I left Dr. Gold's office and ventured out into an unseasonably warm winter afternoon, holding Casey tightly by the leash even though he never ventures far, I couldn't shake one nagging question: Did my dog really want to spend months in a motorhome — with *me*?

PART ONE

1.
IN WHICH CASEY HATES THE RV

If there exists a person in Provincetown, Massachusetts, who dislikes dogs, I have yet to meet him.

Take a summer stroll through this quirky and scenic coastal town on the tip of Cape

Cod, and you will come upon big dogs and little dogs, purebreds and mutts, rascals and pushovers. You'll meet dogs in art galleries, dogs in restaurants, dogs in coffee shops, even dogs in banks. Visitors can take their dogs on whale watching tours or sunset cruises. In 2010, *Dog Fancy* magazine named Provincetown the most dog-friendly place in America.

Though the town's population balloons in the summer months, Provincetown mostly shuts down — and clears out — in the winter. January and February, in particular, can test the resolve of the brave folks who live here year-round, which might explain why so many of them share their life with a dog.

"There's nothing better to keep you company when the loneliness hits," a long-time resident told me at Joe's Coffee & Cafe, one of the few businesses open in the winter. Not far from him, another man's rescue Greyhound lounged regally on a padded bench meant for humans.

I'd moved to Provincetown from Boston in December of 2011, only two months before I was scheduled to embark on my cross-country journey. I hadn't been in the best spirits during that transition. I wasn't fully recovered from the implosion of two meaningful relationships in my life — a romantic one that had ended months before, and a close friendship that had very recently soured.

The first was unquestionably my fault, and it had induced in me the kind of guilt and regret that sometimes made it difficult for me to get off the couch and walk my dog. Instead, I would read a book about walking dogs. *Dog Walks Man,* by John Zeaman, celebrates the simple act of dog walking — "this application of feet to dirt" — in a world consumed by technology and distraction.

"There is a hope that a dog injects into every walk," Zeaman writes. "More than a hope — an expectation, really — that this is going to be something wonderful."

I hadn't been doing much wandering with Casey in the months after the breakup. Instead, I relied on a professional to make up for my failings. For $15 a pop, a dog walker would show up at my front door and save Casey from me. Best of all, she didn't judge me for it. I was easy to judge, lounging around in my pajamas at 11 A.M. on a Tuesday, reading a book about walking dogs. But if the dog walker thought as little of me as I thought of myself those months, she didn't let on.

Before long, Casey had fallen madly in love with her. He would spin in circles and bark wildly when she called his name from the front door. He clearly associated her with one thing — walks! — which led to some confusion when I invited her to a gathering at my apartment. Unsure about why he wasn't being taken for a walk, Casey followed her around for an hour as she tried to reach the cheese dip.

I eventually got my act together and started walking Casey again. In the morning, he and I would stroll down one of Provincetown's beaches. In the late afternoon, I would throw tennis balls for him at a baseball field. In between, I would park myself at Joe's and plan our cross-country adventure.

I should say this: planning does not come easily to me. I'm disorganized, forgetful, and prone to misplacing the piece of paper on which I've laid out the plan. "Oh, let's see where the road takes us," I'm increasingly fond of saying, a trait that I inherited from my mother, a world traveler who despises tourists and will go hundreds of miles out of her way to avoid one.

For this trip, I sought out a middle ground — enough planning that I was sure to meet a variety of American dogs and dog lovers, but enough flexibility that I could wander where my heart and Casey's nose desired. Yes, I had a book to write about dogs in America. But I also wanted to *see* America. I'd yet to visit many of our country's most cherished treasures, including Yellowstone Park and the Great Smoky Mountains. Nor had I strolled through some of our most eccentric cities — Asheville, Savannah, and Santa Fe, to name a few. All were on the itinerary.

Driving an RV in the cold is no fun, so I intended to make like a snowbird and head south for most of February and March. But I had dogs (and humans) to meet along the way. In New York City, I planned to attend the purebred spectacle known as the Westminster Kennel Club Dog Show and visit the city's oldest dog park.

In D.C., I hoped to get lucky and catch a glimpse of the First Dog, Bo. After a couple stops in Virginia (including an afternoon at a dog-friendly winery), Casey and I would hike the Smoky Mountains in North Carolina with a pack of Siberian Huskies and meet a man who lives with wolfdogs.

In Florida, I planned to attend a massive pet industry expo and a conference on animal cognition. Casey, meanwhile, would get plenty of fun in the sun. We'd be entering a dock jumping competition and doing doga — yoga with dogs — on the beach.

From Florida, we would drive to Louisiana and then Texas, where I'd spend a few days in Austin (one of a growing number of "no-kill" shelter communities) before heading to Marfa, a small artsy town in the high desert. Then I'd drive north toward New Mexico, where I wanted to spend a night in the best-named town in America — Truth or Consequences.

In Colorado, I planned to visit a working cattle dog ranch. In Utah, I would tour Best Friends Animal Sanctuary, a home for many unwanted and unadoptable dogs — including several of the Pit Bulls formerly owned by Michael Vick. From there I'd head south through Arizona, where in mid-April Casey and I were scheduled to arrive at my father's desert home. We would relax by the pool for a few days before embarking together on the second half of our journey.

At least, that was the plan.

On the eve of my departure, my friend Dylan dropped by for a visit.

It was a cool, misty, spooky February night in Provincetown; the town was deserted. Dylan and I took Casey for a walk down Commercial Street, the main drag, and in thirty minutes we saw only two signs of life — a town drunk stumbling toward the beach, and a fox darting through a side street.

Dylan was quieter than usual, and I finally asked him what was wrong.

"I have this fear that I'm never going to see you or Casey again," he said.

"What do you mean, never see us again?"

"Oh, I don't know," he mumbled, staring down at his shoes. "You're driving so far, and you're not a very good driver."

"I'm an excellent driver," I protested. I must have sounded like Dustin Hoffman in *Rain Man*.

But Dylan was right — when it comes to oversized vehicles, at least, my track record is poor. Twice I have rented a U-Haul truck to move from one apartment to another, and twice I have sideswiped a parked car as I tried to turn.

"You have to promise me that you'll pay attention to what you're doing," Dylan said, sounding genuinely concerned. Then he handed me a harmonica. "Take this. If you're going to drive around for four months, I figure you could use a har-monica."

Dylan had never been inside an RV, so we spent a few minutes inspecting my rented home on wheels. (I'd picked it up with a friend the previous day at an El Monte RV rental location next to a busy highway in Linden, New Jersey.) My motorhome was twenty-five feet long and had the word "Chalet" printed on its outside, near the main side door.

Though Dylan wasn't a fan of the RV's tacky linoleum floor, he did marvel at the size of the full bed. "You might actually be able to sleep in this," he said. He opened the medium-sized refrigerator, which was paneled with faux wood that matched the motorhome's many cabinets, and then pivoted to face the microwave and miniature gas range and oven. Next to them was a split sink, clean and gleaming.

On one end of the kitchenette was a black panel with buttons and levers whose functions mostly eluded me. I'd been given a brief tutorial at El Monte headquarters, but the woman who'd helped me had mostly laughed off my follow-up questions. Why were there two different water tanks? When was I supposed to fill them? Why did some of the appliances run on gas, while others relied on the auxiliary battery? When was I not allowed to run the generator? Was it when the gas tank valve was open? Closed? And what was the worst that could happen if I screwed this all up? Might the RV explode?

"You'll get the hang of it," she'd assured me, handing me a thick instructional manual and reminding me of the number to call for roadside assistance.

I was hoping Dylan would assuage my fears about traveling for months in an RV with only a dog, but he was little help. "Have I told you that you're crazy to be doing this?" he said, peeking inside the motorhome's cramped bathroom.

He was wise not to step inside — unless you're standing in the shower, there's virtually no space to turn around. If you're tall (I'm six-two), your knees nearly touch the opposite wall when sitting on the toilet. The shower itself is on a platform, and to accommodate taller bathers, the top of the shower was cut out and replaced with a plastic dome that protrudes from the roof.

"I really could use your support," I told Dylan as I sat on one of the two padded polyester benches in the dinette area. "I'm scared to death about doing this alone."

"Oh, it'll be okay," he said unconvincingly. "Besides, you won't be alone. You'll have Casey!"

"But what if Casey hates the RV?"

Dylan shook his head. "He's adaptable —

41

he'll do fine."

I told Dylan about *The Difficulty of Being a Dog,* a beautiful collection of vignettes by French writer Roger Grenier. In one, Grenier compares dogs to German philosopher Immanuel Kant. Dogs are like Kant, who "always want to take the same walk," Grenier wrote. "The less it changes, the happier they are."

"I wonder if dogs are happiest going on new adventures," I said, "or if they prefer structure and predictability."

"It probably depends on the dog," Dylan replied.

As we sat at the dinette, he confessed that he'd hoped my RV would be bigger — maybe one of those thirty-eight-footers with slide-outs, hallways, and multiple living areas. But those motorhomes were expensive, not to mention a challenge to navigate. I needed something I was less likely to crash. Though much of the trip would be spent on the open road, I'd also be driving through some busy metropolitan areas — including Los Angeles, San Francisco, and Chicago.

I'd considered not renting an RV at all and simply driving around the country in my Honda SUV. I'd never crashed *that.* But I didn't want to spend four months in hotel rooms. There was also something appealing to me about life in an RV. One of my fondest childhood memories is of driving from San Francisco to the Grand Canyon in a motorhome with my mom and some family friends. (It was a rare departure from our usual vacations of traipsing through museums and old churches in France, which my ten-year-old self failed to appreciate.)

One night, we'd parked the RV in a lot near the Las Vegas airport. While the grown-ups got drunk on red wine, my friend Odin and I climbed onto the roof and hurled rocks into the sky as planes roared overhead. I remember telling Odin that I wished we could drive around in that motorhome forever.

Looking back, what probably appealed to me most about that road trip was the sense of belonging to a pack. I was an only child, desperate for connection. I didn't mind being crammed in with five other people for a week. I loved it.

When I initially decided to drive around the country with Casey, a friend suggested he come along. Another urged me to bring a camera crew. But while the prospect of human company was tempting, I decided to do most of the trip with only Casey.

For these fifteen weeks, at least, my dog

would be king.

I am not a morning person — a fact that my dog has come to grudgingly accept — and when I awoke at nine inside my Provincetown apartment on Day 1 of our journey, Casey was standing by the foot of my bed and staring intently at my bedhead.

I gazed back at my handsome dog. Casey is built more like a streamlined Greyhound than the typical paunchy yellow Lab. He has a long, narrow, sandy snout. Fine white whiskers protrude from the darkness between his nose and rubbery lips, pricking up when he opens his mouth to yawn. His brown eyes — Casey's darkest feature, and the only evidence of his true age — are weary and inquisitive, and they're accessorized with dainty, humanlike eyelashes. He holds his angular skull low (emphasizing his high shoulders), and his spindly legs support a taut white torso, which fades back into a tucked-up flank.

Before introducing Casey to the Chalet, I took him for one last walk on the beach. It was high tide, but Casey chased his tennis ball like a champ and returned it, as is his custom, some eight feet away from me.

My dog can't accurately be called a troublemaker, but he seems to enjoy this particular power struggle: He drops the ball a few paces away from wherever I happen to be standing. I tell him to bring it closer. He barks. I tell him to stop barking. He barks again. I tell him to cut it out. Disgusted, he pushes the ball a foot toward me with a flick of his nose. I tell him to try again. He barks. I tell him to stop barking. He pushes the ball another foot. I preemptively tell him not to bark. He barks. I tell him he's being a very bad boy. He pushes the ball another foot. And so on.

The ball eventually arrives at my feet, but I could read an entire book on dog training in the process. I probably should have taught him to bring the ball *all the way* to me, but it felt somehow wrong to break him of one of his few problem behaviors — his rare acts of rebellion. Besides, I engage in his little game only when I'm feeling obstinate. Usually I just give in, walk a few paces, and pick up the ball. Or I walk in the opposite direction, forcing Casey to gather the ball and hurry up after me. Either way, Casey's winning — because I'm walking.

I let him win a few times on that last morning in Provincetown. Then, at about noon, I opened the RV's side door and tried to sweet-talk Casey inside.

"Go on," I said. "It'll be fun!" He paced

instead around the motorhome, found a nearby bush, and peed.

When I tried again a minute later, he galloped over to the closed rear door of my Honda, as if waiting for me to let him leap into his usual spot in back. Finally, I climbed inside the RV and called for him. Casey came to the open side door and looked at me with apprehension, but what other choice did he have? He likes to be where the humans are. He gave in and jumped in.

Then he ate my roast beef sandwich. Casey snagged it off the driver's seat while I was organizing my clothes in a cupboard above the bed. "Way to make yourself at home," I said, waving the empty sandwich wrapper above my head.

Casey's shoulders dropped, and his eyes drifted toward the floor. He looked guilty as charged. Or was I anthropomorphizing? The question of whether dogs are capable of feeling guilt — or whether they're reacting instead to a raised voice, or a split-second change in our body language — is one of many questions being debated by canine cognition researchers across the globe. Most believe that while dogs do understand when they've broken a rule, they lack the ability to feel remorse.

But try telling that to a dog owner who

comes home to find the garbage overturned and the dog slumped in a corner. The dog certainly *looks* guilty. A few days before my departure, I'd read about a Basset Hound who was suspected of eating a $4,500 diamond wedding ring.

"She was the only one in our room, so we immediately looked at her — and she looked guilty," Rachel Atkinson told a New Mexico television station. (Basset Hounds are prone to swallowing things they shouldn't. In 2010, one in Colorado Springs helped herself to thirty-one nails and her rabies tag.)

Though Atkinson was convinced that her dog "looked guilty," a study by psychologist and animal cognition researcher Alexandra Horowitz casts doubt on our ability to effectively label what our dogs are thinking. Horowitz asked study participants to tell their dog not to eat a treat, and then to leave the room. A researcher would then either give the dog the treat or throw it away. Upon returning, owners were told (often incorrectly) that their dog had either resisted temptation — or defied them.

Owners who thought their dog had disobeyed perceived the dog to look guilty, even if the dog wasn't. In fact, innocent dogs who were nonetheless scolded were thought to look *more* guilty than dogs who'd actually eaten the treat.

Whether Casey felt guilty or not, my sandwich was still inside his stomach. I dispatched a friend to get me a replacement while I finished getting ready. The plan was to make the 115-mile drive to Boston and take Casey for a walk in Franklin Park, where he'd been first socialized with other dogs as a puppy.

Then I would drive to Cambridge and meet Amy Hempel, a dog-obsessed fiction writer who coedited (with Jim Shepard) the 1995 collection *Unleashed: Poems by Writers' Dogs.* A delightful exercise in anthropomorphism (conceived of after many cocktails), the collection imagines what dogs might say if given the power of poetic license. One of my favorite poems is by Karen Shepard, who wrote about her dog, Birch, though she could have been writing about most dogs:

You gonna eat that?
You gonna eat that?
You gonna eat that?
I'll eat that.

I had no idea where Casey was going to position himself while I drove to Boston.

My bed seemed a natural choice, but that was all the way in the back of the RV. In case that didn't suit him, I'd placed his doggie bed under the table and between the padded benches.

"These are both terrific options," I said, petting him on the top of his bony head.

By 1:30, I was ready to go. Casey jumped onto the bed, where I hoped he'd stay, and I positioned myself in the driver's seat. I took a deep breath and remembered what an uncharacteristically helpful tollbooth operator had told me as I drove over the George Washington Bridge the previous day on my way to Provincetown from El Monte's New Jersey office.

"Go as slow as you need to go," he suggested. "The assholes can go around you."

There had been two assholes during that first terrifying drive from New Jersey: a trucker cut me off and nearly sent me barreling into Long Island Sound, and a middle-aged woman gave me the finger when I misjudged the distance of her car in my passenger-side mirror. But I'd survived, and those seven hours alone in the RV had given me a strange kind of confidence.

"I can do this," I thought, starting the engine for Day 1 of our voyage.

Casey did not like the loud, deep, guttural sound of the Chalet's engine. "Really?" I said, as he sprang off the bed and came to stand next to me. "We're not even moving yet."

I gave Casey some comforting pats, unbuckled my seat belt, and walked him back to the bed. I kept the engine running, in hopes that he merely needed to get used to it.

"Up you go," I said. He looked at me with his sad face. "The bed is really quite nice," I insisted. He didn't move. "Good thing I keep you thin," I told him, lifting him into my arms and placing him on my bed. Then I said "Down," a command that Casey understands to mean two different things: lie down, or jump down from whatever surface he happens to be on.

Casey's not dumb — he chose the latter interpretation and tried to leap off the bed. I was a step ahead of him, though, and blocked his path with my hip. "Just give the bed a try," I pleaded. Then I said "Down" again, this time elongating the word — dooooooowwwwwwwnnnnnnnn — and repeating it twice, as I sometimes have to do when he's being stubborn. He finally lay down.

"Good boy!" I said, more enthusiastically than usual.

I walked back to the driver's seat. "Now, stayyyyyyyyyyyyyyy," I told him as I buckled my seat belt again. He's normally quite good at staying. For one of our favorite outdoor games, I tell him to stay and then walk about a hundred feet away. He waits patiently until I scream "Okay!" and then sprints excitedly to me as if he hasn't seen me in weeks.

I put the gear shift in drive and tapped lightly on the accelerator, causing the RV to move several feet. Casey's head cocked up, and his body stiffened. I pressed on the accelerator again and began driving slowly around the outdoor parking lot where I'd loaded up the RV. Casey jumped up and off the bed again and went to stand awkwardly by the side door. He wanted out.

My heart sank. I'd read in several online RV forums that dogs can take time getting used to a moving motorhome, and I imagined the worst. What if Casey never got comfortable? What if this cross-country journey designed partly to help me bond with my dog became just another thing for me to beat myself up about? And how much would he have to hate life on the road before I called off the trip entirely?

I stopped the RV, unbuckled my seat belt, knelt down next to Casey, and kissed him on his wet nose. "Buddy, it's going to be okay," I said. I think I was trying to convince the both of us. We stayed like that for a minute before I led him over to his doggie bed under the dinette. I figured a small space might make him feel more secure. "Dooooowwwwwwnnnnnnnnn," I said again. He sighed in resignation.

Casey didn't like the doggie bed, either, so he spent most of the drive to Boston seated awkwardly in the small space between the driver and passenger seats. I took my mind off Casey's predicament by making phone calls. One was to a friend, Peter, who happened to be going through his own dog-related drama.

"My wife's using our dog as a weapon against me!" he blurted out before I could tell him that I'd just embarked on my cross-country adventure, and that this conversation really needed to be about me.

Technically, the woman Peter was referring to was his *ex*-wife. They'd only recently divorced, and Peter wasn't sure if he'd stayed in the three-year marriage to honor his wedding vows or to avoid losing the dog.

They'd married too quickly, Peter realized in retrospect, and a year into their relationship it had become clear to him that they had little in common except for the young Pit Bull/German Shepherd mix they'd adopted together.

Though the dog had served as an effective distraction from their marital problems, it had also sparked both spoken and unspoken resentments. Peter couldn't stand that the dog seemed to prefer his wife — and that it slept on her side of the bed. She complained that Peter was more affectionate with the dog than he was with her, a claim that Peter denied even though he knew it was true. (Had I known earlier about Dr. Gold's *When Pets Come Between Partners,* I would have bought Peter a copy.)

Peter couldn't imagine life without his dog. "She knows he's what I really care about, and she resents it," he told me as I rolled past the town of Orleans on Cape Cod. "I'd be happy with joint custody, but I don't know if she'll ever be willing to share him."

And if Peter raised the custody issue in court, the odds would be stacked against him. Judges in animal disputes tend to rule in favor of women, just as they do in child custody disputes. Though pets are still technically considered property in the eyes of the law ("a dog is the legal equivalent of a sofa," Stanley Coren writes in *The Modern Dog*), most judges now understand dogs and cats to be far more than that — a "child substitute," as one judge put it.

In deciding pet custody cases, many judges seek to identify which human the dog is most bonded to. "There's definitely a trend in recent years where we see judges really looking out for the best interest of the pet," Jennifer Edwards of the Animal Law Center in Denver told me. "They're applying the same standards they would with a child." (In one unusual custody dispute over a mixed-breed dog in St. Louis, the judge told the divorcing couple to stand on opposite sides of his courtroom and call for the dog. But the test failed miserably, as the confused animal shuffled over to the judge instead.)

Peter wasn't about to put his dog — or himself — through that, and he had no plans to take his ex-wife to court. "I just hope she'll stop hating me long enough to realize that I deserve to see my dog," he told me.

"Just give her some time," I replied. "She'll let you see him."

"I hope you're right," he said, sounding

more depressed than I'd ever heard him.

It was nearly dusk when Casey and I arrived in Boston. I carefully navigated rush-hour traffic and parked the Chalet on a mostly empty street a few blocks from the entrance to Franklin Park in Jamaica Plain, the neighborhood where I'd lived with Casey for most of his life. Casey seemed relieved to be stopped, and he wagged his tail expectantly when I said the magic words — *Wanna go for a walk?*
Jamaica Plain (known to residents as JP) is a dog walker's delight. Once dubbed the "Eden of America," there's green around virtually every corner. Portions of the Arnold Arboretum — the oldest public arboretum in the country — are so beautiful that you'd be a fool not to take a promising first date there. Nearby is Olmsted Park, part of Boston's Emerald Necklace, which borders the sixty-eight-acre Jamaica Pond, the largest body of freshwater in the city. Though dogs technically aren't allowed in the pond, it's a rule that's enforced about as often as the old Massachusetts law that forbids frightening a pigeon.
But my favorite JP haunt is Franklin Park. At 527 acres, it's Boston's largest, with miles of wooded dirt paths that make you

forget you're in a city. Dogs aren't technically allowed off-leash there, but the lack of
a consistent ranger presence along the trails
makes it a no-brainer for dog lovers intent
on exercising their pets.

Unlike fenced-in dog parks, Franklin Park
is a multiuse space, meaning that dogs
encounter a colorful cast of characters on
the secluded paths. There are homeless
people, Tai Chi enthusiasts, high school
cross-country teams, bird watchers, and,
though you rarely see them, practitioners of
the Santeria religion. (Dogs occasionally
lead their humans to a makeshift altar and
the remnants of a sacrificed chicken.)

The dog people tend to stick together,
roaming the park in packs and forming the
kind of vibrant and close-knit community
that makes dog parks "a kind of victory over
the anonymity and transience of life," as
writer Mary Battiata once put it.

When I'd first arrived at Franklin Park
when Casey was a puppy, I was taken aback
to learn that the community didn't just get
together every morning and evening to walk
their dogs — many of its regulars also hung
out with each other *outside the park.* There
were long-lasting friendships and longer-
lasting feuds. There was a phone list. And
there were self-anointed leaders who were

deeply invested in everything from the minutiae of the park's cleanup committees to whether your dog was getting enough exercise.

The queen bee of Franklin Park's dog lovers was a short, stout woman in her late fifties named Gerry. She ran a small pet-sitting operation out of her apartment but spent much of her days at the park walking dogs and lecturing humans.

"Your dog's a little overweight," she announced not long after meeting me. "What are you feeding him?"

I recited the name of a brand suggested to me by my vet. "Well, you certainly can feed him that if you want to *kill him,*" she barked. Gerry then proceeded to lecture me about the evils of mainstream dog food, insisting that I trek across town to a small pet supply store that carried the two or three obscure brands she deemed acceptable.

Most of us worked hard to get on Gerry's good side. Once she liked you and your dog (she usually liked your dog well before she liked you), she was downright sweet. There was also no denying that she knew what she was talking about when it came to canine temperament and health. Her dogs were models of civility and slenderness. Plus, they seemed to live forever. Her friendly mutt,

Buddy, lived to be sixteen and walked with her in Franklin Park until the end.

I was hoping to run into Gerry — or any of the other park regulars — but the trails were mostly deserted. Still, it was comforting to begin this journey with a familiar walk. Casey bounded ahead of me on the dirt path, and in the woods he stopped and waited for me in front of a tall, jagged rock. For as long as I can remember, I've placed a tennis ball on that rock each time we've walked past it.

On my signal, Casey stood tall on his hind legs and stretched to gently grab the ball in his mouth. Then he ambled along, proud as could be.

After nearly an hour in Franklin Park, Casey and I climbed back into the motorhome and headed to Cambridge to pick up Amy Hempel, who teaches creative writing at Harvard. It's fitting that Amy's office is located in a building called the Barker Center. Amy's fiction is overrun with animals, especially dogs.

In her novella *Tumble Home,* the narrator is tortured by the memory of her mother giving away her dog as a child. In Amy's short story collection *The Dog of the Marriage,* she writes about a woman who trains guide dogs for the blind, and about another who hires a pet psychic to help her find a lost dog. In "In the Animal Shelter," a short piece in *The Collected Stories of Amy Hempel,* Amy writes about women who are spurned by men — and who seek emotional revenge at animal shelters by doting over, but ultimately rejecting, homeless dogs and cats.

"And I know where they go, these women, with their tired beauty that someone doesn't want," Amy writes. "They reach out to the animals, day after day smoothing fur inside a cage, saying, 'How is Mama's baby? Is Mama's baby lonesome?' . . . But there is seldom an adoption; it matters that the women have someone to leave."

At the time of my visit, Amy had recently completed another short story about life inside an animal shelter. "A Full-Service Shelter," published in *Tin House,* is labeled fiction but is based on Amy's heartbreaking experiences volunteering at a high-kill animal shelter (a "house of horrors," Amy calls it). She writes about helpless, traumatized dogs; about the Sisyphean devotion of volunteers who take Death Row dogs — mostly Pit Bulls — on their last walks; about underpaid shelter employees who seem inexorably calloused by the often monstrous nature of their jobs; and about the shelter's uncaring, incompetent leadership.

As I sat with Amy on the carpeted floor in her home's empty attic — a space that doubles as her doggie play area — she told me that she used to write short, poignant biographies of each shelter dog for rescue organization websites. Combined with a flattering picture, a well-crafted biography "can mean the difference between life and death for these dogs," Amy said as Casey and her Lab, Wanita, wrestled around us.

Maybe it was because Amy and I were in her dark, cavernous attic, as if about to begin a séance, but she looked like a beautiful witch. Her long gray hair glowed silver and gold against the shadowy attic walls.

"I'm a writer, and I would put everything I could into those bios," Amy continued. "But the trap for me was that I would say, 'Hey, I've got three hours. I could use it to write three sentences of a short story that I'll probably throw out, or I could save a good dog's life or a few good dogs' lives.' When I thought about it that way, I realized that I might never write fiction again."

Amy hasn't stopped writing, but she devotes much of her energy these days to dogs. She's a founding board member of the Deja Foundation, which offers assistance to rescue organizations working to remove dogs from shelter euthanasia lists.

"Where writing used to be, advocating for shelter dogs and Pit Bulls is now," she told me. "If I ask myself if I get more happiness from writing or from dogs, it's not even close."

Amy used to be involved with Guiding Eyes for the Blind. Wanita, who is nine, has birthed four litters and thirty-three puppies, many of which work as guide dogs today. "Two even walked a blind couple down the aisle at their wedding!" Amy told me, sounding like a proud grandmother.

As Amy gushed, Wanita stretched her neck forward to kiss her human. Amy puckered up and obliged, stroking Wanita's face with her long, delicate fingers. "You did good! You made good guide dogs, Wan," Amy said, tilting her head back with her eyes closed, seemingly overcome with delight.

I had a sudden urge to pet Casey, but he was nowhere to be found. After nearly twenty minutes chasing Wanita around the attic, he'd apparently given himself a time-out. He reappeared a few minutes later, his white, fluffy tail wagging as he shuffled toward us.

"What a beautiful dog," Amy said, studying Casey's long face.

We spent most of the next few minutes in silence. There is arguably no better cure for stress than lounging around on a floor with some happy Labs, and I could feel my body relaxing for the first time all day. Amy and I had bonded almost instantly. It wasn't, I suspected, because we're both writers. We'd bonded over something more visceral: we're both dog people.

In Amy's attic, it occurred to me that my journey would be filled with moments like these — quiet moments with fellow dog lovers who instinctively know that when you're in the presence of dogs, there's no need to strain to fill the silence. And though I worried that Casey might never come to love the RV, I realized that at each stop he would at least be greeted by humans and dogs wanting nothing more than to make his day.

"Maybe Casey won't hate this trip after all," I told Amy later that night at a Chinese restaurant, having already confessed to her my trip-related insecurities. Amy insisted on picking up the tab (when you're driving around the country in a motorhome, everyone wants to feed you), and after dinner we returned to her house to find Casey resting on her couch. I hugged Amy and Wanita good-bye, lured Casey to his feet with more magic words ("Wanna go outside!"), and drove the RV to my friend Jay's place in Waltham, a suburb of Boston.

Jay had agreed to let me park the motorhome in a lot next to his building, which isn't far from the studio where Aerosmith recorded in the late 1970s. After showering in his apartment (I'd been warned not to put any water in the Chalet until I arrived in Florida — something about freezing pipes), I returned to the RV to find Casey curled up on the bed next to the football-shaped teddy bear I'd purchased on a whim a few weeks earlier.

I was exhausted and eager to join Casey. I'm one of the roughly half of all American dog owners who allow their dog on the bed, though sometimes I have to work to get him there. Casey isn't one for long cuddle sessions, but he'll humor me for a few minutes before either moving to the other side of the bed or leaving it entirely. Casey may have been better matched with a married man over forty-five, the demographic group least likely to want to share their bed with a dog.

Just as I was about to fall asleep, I received a phone call from an old friend I hadn't spoken to in nearly a year.

"Anything new?" he wanted to know, eager to catch up.

"As a matter of fact . . ." I said, explaining that I was on the first day of an epic cross-country journey with my dog.

"Oh, like Steinbeck's *Travels with Charley*?" he wondered.

I was getting used to that response. "Yes," I replied. "Except, about dogs."

There was another important difference, though I hesitated to bring it up. Compelling evidence suggests that *Travels with Charley* isn't entirely — or even mostly — nonfiction. In 2010, journalist Bill Steigerwald retraced the author's ten-thousand-mile journey and concluded in a *Reason* magazine article (titled, "Sorry, Charley") that "the iconic American road book was not only heavily fictionalized; it was something of a fraud."

Though Steinbeck claimed to have roughed it in a camper he named Rocinante (after Don Quixote's horse), Steigerwald discovered that the author had mostly stayed with friends or in fancy hotels. Many of the people Steinbeck claimed to have met on the road seemed to be invented, and the author didn't visit many of the places he'd claimed to.

Steigerwald doesn't believe that Steinbeck intended to fictionalize his book. "He was desperate," Steigerwalk wrote. "He had a book to make up about a failed road trip, and he had taken virtually no notes . . . As he struggled to write *Charley,* his journalistic failures forced him to be a novelist again." (Steinbeck's son, John Jr., was less generous. "He just sat in his camper and wrote all that shit," he told Charles McGrath of *The New York Times.*)

Does that make *Travels with Charley,* which enjoyed a year-long run on *The New York Times* bestseller list and is still deeply revered, a bad book? Certainly not. Even those who think it was mostly fiction find themselves charmed by it.

"In many ways it is still a wonderful, quirky, and entertaining book," Steigerwald concluded in *Reason.* (He went on to write a book about his discovery called *Dogging Steinbeck.*)

Bill Barich, who wrote *Long Way Home: On the Trail of Steinbeck's America,* believes Steinbeck "made up most of the book" but still admires the author's cultural analysis. "I still take seriously a lot of what he said about the country," Barich told McGrath. "His perceptions were right on the money about the death of localism, the growing homogeneity of America, the trashing of the environment. He was prescient about all that."

I found much to like about the book, too. Steinbeck's descriptions of Charley, in particular, are delicious. "Charley is a born diplomat," Steinbeck wrote in his introduction. "He prefers negotiating to fighting, and properly so, since he is very bad at fighting. Only once in his ten years has he been in trouble — when he met a dog who refused to negotiate. Charley lost a piece of his right ear that time."

A prolific anthropomorphizer, Steinbeck noted that Charley "would rather travel about than anything he can imagine." I was envious of Steinbeck's seeming certainty. How nice it must be, I thought, to know that your dog could dream up no better way to spend a few months.

I couldn't relate to that, but I took to heart Steinbeck's travel planning advice. A meaningful adventure, Steinbeck believed, rarely sticks to the plan.

"A journey is a person in itself; no two are alike," he wrote. "And all plans, safeguards, policing, and coercion are fruitless. We find after years of struggle that we do not take a trip; a trip takes us."

I was eager to be taken.

The next morning, I awoke to a man shouting profanities at his dog.

"Dammit, Jasper, stop eating the garbage!"

I opened the drape next to my bed to find an old man in a New England Patriots sweatshirt shooing his dog away from a discarded pizza box. I felt for the guy — Casey can sniff out a rogue slice of leftover pizza from blocks away. He'll also sometimes help himself to a steaming hot pie right out of the box.

In an especially brazen maneuver, he once snatched a slice of pepperoni from a teenage couple devouring a large pizza in San Francisco's Golden Gate Park. I'd apologized profusely and offered to buy them another pie, but instead they gave Casey a second piece.

"Dude, I totally get it — we have the munchies, too," the boy said, before offering me a joint.

Though the forecast for my journey's second day called for a possible storm, the morning skies looked unthreatening. I took Casey for a walk, scarfed down a bowl of cornflakes, and turned on the RV's engine. Casey made a sad face, but I was determined not to let his attitude get me down. I was excited to be heading to beautiful western Massachusetts, where Casey was going to pose for Amanda Jones, one of the country's leading studio pet photographers.

I've been obsessed with photographs of dogs since coming upon the work of Elliott Erwitt when I was a teenager. The author of four books of dog photography (including *Son of Bitch*), Erwitt captures dogs in playful, ironic, often humanlike poses. He's perhaps best known for his 1974 image "Felix, Gladys, and Rover," which shows a woman's booted legs between a Great Dane's and a Chihuahua. But my favorite Erwitt photograph (titled *Dog Show*) depicts a white Standard Poodle on its hind legs leaning against a low railing at a dog show in England, checking out the action.

I later discovered the work of William Wegman, probably America's most famous pet photographer. He's best known for his pictures of Weimaraners and frequently portrayed his most famous canine subjects — Man Ray and Fay Ray — as humans. In 1982, *The Village Voice* named Man Ray "Man of the Year."

Wegman's success aside, *The New York Times* noted in 1990 that pet photography "ranks just above baby photography among least-favored photographic specialties." Part of the problem, as the *Times* had pointed out a century earlier, is that "it is a difficult matter to get . . . dogs in a picturesque position."

But that's also part of the fun. And as Americans have turned to the Internet and social media to chronicle (some might say *advertise*) their lives through pictures, dogs and cats have taken a starring role. Facebook and Instagram have turned average dogs into worldwide celebrities. Countless people have created Facebook and Twitter accounts for their dogs. Boo, a Pomeranian from California, has more than eight million fans on his Facebook page, which mostly features cell phone pictures taken by his owner. (Dogs are probably the least likely Facebook "members" to get defriended.)

As amateur pet photography has exploded in popularity, so, too, has professional pet portraiture. Search for "pet photographer" today in any major American city, and you're likely to be overwhelmed with options. Some will photograph your pet in a studio. Others will come to your home, and still others specialize in capturing your dog outdoors. But few are as respected as the woman I was driving to meet.

On my way to Amanda's studio in North Adams, I drove along the historic Mohawk Trail. In the summer and fall, this scenic former Native American migratory game path is packed with visitors, many of them traveling by motorhome. On this cold and clear Saturday in February, though, I saw only a handful of cars as I rolled through the 6,400-acre Mohawk Trail State Forest, home to deep gorges and five-hundred-year-old eastern hemlock trees. The tallest tree in New England — a 171-foot eastern white pine — is also here, though its exact location is a secret.

It had snowed recently, and there were patches of white along the windy, two-lane road. As I approached North Adams, the trail dropped some seven hundred feet in two miles, necessitating an intimidating hairpin turn. If your brakes give out here, or if you're distracted by the stunning views of the northern Berkshires, you might drive straight into the Golden Eagle, a restaurant that sits close to the road.

As I carefully navigated the turn, I saw two black Labs standing side by side in the Golden Eagle parking lot. They were waiting for their owner, a pint-sized woman with dyed red hair, smoking a cigarette as she gazed down the Hoosac Range.

Her dogs reminded me of an improbable Mohawk Trail canine survival story. In 2009, an Ohio couple, Kathy and John

Dunbar, were traveling to meet a family member in Maine when they stopped along one of the trail's many turnouts to let their two dogs relieve themselves. But when Kathy and John got back in the car, each assumed that the other had grabbed the younger dog, a Maltese Poodle mix named Mindy. It wasn't until forty-five minutes later that they realized their mistake. They sped back and searched the area for hours, but Mindy — whom they'd nicknamed the "ghost" for her ability to surreptitiously enter or exit a room — was gone.

Three months later, a father and his son were hiking nearby along the Connecticut River when they spotted a small, emaciated dog stumbling in circles, blinded by a mat of fur that had grown over her eyes. It was Mindy. She had lost half her body weight, but she'd somehow survived in woods populated by coyotes and bears. The dog was safely returned to Ohio, where she's living as a Maltese probably should — *indoors.*

Nestled near the Vermont and New York borders, North Adams is a charming, rough-edged little city in the middle of a mountain range. Once a thriving mill town, it has tried to reinvent itself in recent years as a center of art and culture. It's now home to the country's largest contemporary art museum, as well as hundreds of working artists' lofts.

Casey took an immediate liking to Amanda Jones's airy, 1,800-square-foot studio, located on the second floor of a brick building on the museum's campus. The hardwood floor was littered with tennis balls, chew toys, dog beds, and water bowls, and within minutes of our arrival Amanda was feeding Casey Pup-Peroni beef-flavored dog treats.

My dog will gladly work for food, and he practically galloped over to the white seamless background paper where Amanda — a tall, attractive, friendly blonde in her forties — needed him to stand. Though some dogs dislike bright photography lights, Casey barely seemed to notice them. Instead, he kept his eager brown eyes on Amanda, who gripped treats in her right hand while she knelt on a black pad, cradling her digital camera against her chest with an underhand grip.

To get Casey to turn his head to a profile position, she would swivel her right arm to the left or right, or pretend to throw a treat toward a brick wall adorned with dozens of Amanda's cover photographs for *The Bark* magazine. Casey — tail wagging, smile on his face — proved to be a model model.

"He's a total ham!" Amanda's assistant said.

"Oh my God, he's so easy to work with!" Amanda gushed.

I smiled. "I bet you say that to all the dogs."

Amanda laughed. "Actually, some dogs really get spooked," she said. "They'll get scared by the light, or the sound of the camera. When a dog's afraid, then he doesn't want to eat, and that's a big problem because that's how I move them onto the set. If a dog isn't food-focused, this isn't easy."

During a break a few minutes later, Amanda explained that there were only a handful of people known nationally for pet portraiture when she began photographing dogs in 1994. "One of my goals was to up the quality of animal photography, because a lot of it was pretty cheesy," she told me. "Puppies in baskets, that kind of thing. I would never, ever, put a hat on a dog! I'm pretty adamant that dogs are dogs — they aren't humans. They're beautiful and striking enough without humanlike accessories."

Amanda walked me over to a computer and showed me the photos she'd taken of Casey to that point. Casey looked great, to be sure, but Amanda had also managed to capture the many elements of his personality — the playfulness, the eagerness to please, the laser focus on what might be happening next, and even the apparent sadness and frustration that I'd spent years interpreting as an indictment against me.

What did it mean that Casey occasionally appeared dejected in Amanda's studio, a place oozing with love, chew toys, and food? Was Casey actually sad? If he seemed happy in ten pictures and sad in one, did the one matter? And what might he be sad about? Was he disappointed that Amanda hadn't given him a treat in thirty seconds, or was his wound deeper, more significant?

There was also the possibility that I was misinterpreting Casey's facial expressions and body language. Dogs make about one hundred different facial expressions, including ear movements. Children are especially prone to misjudging them; in one study, nearly 70 percent of four-year-olds interpreted an aggressive dog as smiling and happy.

But adults don't always get it right, either. A dog's yawn, for example, is often perceived by humans to mean a dog is tired, or bored. But more often, a dog yawns to calm itself in a stressful situation — when a kid jumps on him or when someone approaches in an aggressive way. It means the dog's uncomfortable, or would rather be left alone.

(Sometimes, though, dogs yawn for the same reason humans yawn — because yawns are contagious. In humans, contagious yawning is seen as a sign of empathy. Researchers disagree on whether contagious yawning is a sign of empathy in dogs.)

In *The Other End of the Leash,* Patricia McConnell writes that humans and dogs are "each speaking our own native 'language,' and a lot gets lost in translation." She notes that we often assume dogs like the same things we do. Take the hug, for instance.

"The tendency to want to hug something that we love or care for is overwhelmingly strong," she writes. But most dogs aren't huggers. "Your own dog may benevolently put up with it, but I've seen hundreds of dogs who growled or bit when someone hugged them. . . . Humans and dogs often miscommunicate, and the consequences range from mildly irritating to life-threatening."

I was amazed at how people misjudged Casey's facial expressions in photographs. When I posted pictures on Facebook of Casey in the RV (a place he didn't like), someone would invariably write, "Look at him, so happy to be traveling. Not a care in the world!"

We tend to see what we want to see in pictures, whether the subject is a dog or a human. A friend of mine once looked at a photograph of me smiling with some classmates in high school. "Oh, look at how happy you are there," she said. But she couldn't have been more wrong. "Look at it again," I told her. "That's a fake smile."

Dogs presumably don't put on fake smiles, but they do send out hundreds of subtle signals that we often miss. Amanda has become an expert over the years at judging a dog's mood, and she'll stop a studio session if she senses that a dog is anxious or uncomfortable.

"Sometimes it's a tough situation, because the dog's owner is pushing you to keep shooting, but the dog clearly isn't having fun," she told me. "And if the dog isn't having fun, then I don't want to be doing this."

I couldn't imagine a dog not having fun with Amanda. But two months later, when Casey and I would see her again in San Francisco, we would all be faced with just such a scenario: a dog of mine, inexplicably camera shy.

■ ■ ■ ■

Though I would have liked to spend the rest of the weekend poking around western Massachusetts, I needed to get to New York City for the Westminster Dog Show.

Normally the drive is a straight shot down the Taconic State Parkway, but as is the case with many pretty roads in the Northeast, RVs are prohibited.

As I was planning an alternate course, I remembered that I had lunch plans in Westport, Connecticut, with a man who'd spent the previous six weeks looking for his lost dog. It's easy to forget about Connecticut — writer Rick Moody once dismissed it as a place you speed through "on the way to somewhere else, somewhere better."

For Moody, the state's saving grace is the Merritt Parkway, which he calls "the finest roadway in the land." The Merritt is thirty-seven miles long and is renowned for its beautiful forestry, creatively designed overpasses, and quaint signage. But I wasn't allowed to drive on it, either. In the end I grudgingly took Interstate 91, the state's busy north–south corridor that runs through Hartford, the state capital. Though it wasn't much to look at from the interstate, Hartford was a favorite of Mark Twain, who wrote, "Of all the beautiful towns it has been my fortune to see, this is the chief."

I thought about stopping twenty miles south in the town of Meriden, but I was running late and am afraid of ghosts. According to legend, Meriden's Hanging Hills are home to a Black Dog that haunts this stretch of trap rock ridges casting their jagged shadows across south-central Connecticut.

Locals attribute at least six deaths to an encounter with the phantom pooch, the most recent being on Thanksgiving Day 1972. As the legend goes, *If a man shall meet the Black Dog once, it shall be for joy; and if twice, it shall be for sorrow; and the third time, he shall die.*

The most convincing account of an encounter with the Black Dog came from nineteenth-century geologist W. H. C. Pynchon. In 1891, he undertook an expedition to West Peak for the purposes of collecting samples of hardened lava. As Pynchon worked, he became aware of the presence of a medium-sized black dog. With its wagging tail and "sad, spaniel-like eyes," Pynchon's new friend was "happy-go-lucky" and even playful, Pynchon later recounted in an article for *Connecticut Quarterly.* For the rest of that day, the dog remained his loyal and amicable companion. That is, until dusk, when it slowly made his way to the trees and "quietly vanished into the woods."

If a man shall meet the Black Dog once . . .

Pynchon encountered the dog again several years later, while he was traveling with a fellow geologist named Herbert Marshall. As the pair rested one evening, their conversation turned to the dog legend. Marshall told Pynchon that he'd seen the dog twice himself, but that he "did not believe in omens unless their were lucky ones." The next day, as they made their way through a deep ravine, the words of the twenty-third Psalm came to Pynchon's head: "Yea, though I walk through the valley of the Shadow of Death . . ." As the two continued to climb, they came upon the Black Dog standing on rocks above them. Marshall turned white.

"I did not believe it before," he quietly said to Pynchon. "I believe it now."

Just then, according to Pynchon, "the fragment of rock on which [Marshall] stood slipped. There was a cry, a rattle of other fragments falling — and I stood alone." Marshall's lifeless body was recovered later by rescuers at the bottom of the ravine. They described the scene as empty, save for the presence of a black dog that seemed to be watching over the dead body and fled as they approached.

I kept on trucking down I-91 and pulled over for gas near New Haven, home to Yale University.

The school had been in the news recently for its decision to rent out a hypoallergenic Border Terrier mix named Monty to stressed-out law students. Yale professor Blair Kauffman told the *Yale Daily News* that Monty was "extremely well qualified" and had graduated "summa cum laude" from his therapy dog course. Yale officials had at first worried that the rent-a-dog program would make the school seem "foolish," but it turned out to be a huge hit. Several students said the dog reminded them of home, and there was a long waiting list for the chance to spend thirty minutes playing with Monty.

As popular as he was, there was little chance Monty could dethrone Yale's top dog — its beloved bulldog mascot, Handsome Dan XVII. Dogs are the most common live animal mascot, probably because they're the only species that we regularly divide into personality types, allowing us a plethora of anthropomorphic possibilities. Dogs are also cheap. By contrast Louisiana State University boosters spent $4 million on the fifteen-thousand-square-foot habitat for their prized tiger, Mike, causing some to wonder if the cash might have been better spent on, say, scholarships for needy students.

I'd met Handsome Dan (nicknamed Sherman) the previous year, when he was the "guest speaker" at Yale's Trumbull residential college. His human, an investment manager and Yale graduate named Chris Getman, had offered to let me tag along.

My most vivid memory of that day was the sight of Sherman — a brown and white bulldog with a large, droopy mouth — defecating on a cement walkway outside Trumbull, a picturesque granite building with gothic arches, Jacobean chimneys, and, ironically enough, a "Potty Court" featuring a statue of a man sitting on a toilet. (It's a whimsical rendition of Auguste Rodin's *The Thinker.*)

The walkway wasn't the most discreet place to go to the bathroom, and when Sherman chose it I wondered if he was punishing us for taking him outside in the rain. Sherman hates water, and he'd endured a full five minutes of it on our walk from the car to Trumbull. But Getman assured me that his dog wasn't smart enough,

81

or "catlike" enough, to orchestrate such re-taliation.

"He's not the sharpest knife in the drawer," Chris told me, though that can safely be said of most Bulldogs, who rank seventy-seventh out of seventy-nine in dog intelligence, according to psychologist and dog intelligence expert Stanley Coren. Only the Basenji and the Afghan Hound are dumber.

As Chris fumbled through his coat pocket for a plastic bag, his wife looked at the bright side. "At least he didn't do it *inside,*" she said.

Among the breed's many health problems, Bulldogs have hypersensitive stomachs and are the most relentless farters in the canine world. But gassy mascots also have their upside. Handsome Dan XIII, who served from 1984 until 1995, famously pooped on the commandant's lawn at Army and threw up after chasing the Princeton Tiger. Both incidents were promptly anthropomor-phized as proof that the dog knew the enemy when he saw it.

Sherman was in his third season as Yale's mascot when I visited, and he had taken to his job "with great enthusiasm," Chris told me. He loved people, loved crowds. Inside Trumbull House, Yale students practically hurled themselves at the dog, who eagerly licked their faces in return. When it was time to get started, Chris sat on a red antique chair next to a black grand piano and regaled the students with the history of Handsome Dan, believed to be the country's first live mascot. Sherman relaxed on the carpet next to him, occasionally poking his head into a plastic bag of treats.

Four of the last five Handsome Dans have belonged to Chris, a longtime dog lover who illegally kept a sheepdog in his dorm room during his senior year at Yale. Proud of his school's mascot tradition, Chris enjoys poking fun at his rival for mascot supremacy — Georgia's Bulldog mascot, Uga (pronounced UGH-uh).

"As you probably know," he told the Yale students, "Georgia has a mascot named *Uga.*" (Whenever Chris says the word "Uga," it sounds as if he's been punched in the stomach.) "Uga is pretty high-maintenance. He flies first class to all the away games. He lives in an air-conditioned doghouse. I don't think he's ever been touched by an undergraduate. He certainly hasn't been walked by an undergraduate! Handsome Dan is a very different kind of mascot. He's a mascot *for the people.*"

I arrived on time for my lunch appointment in Westport, a town of 26,000 in the southwest corner of the state. I'd come to meet Michael Ghiggeri, who had spent weeks looking for his lost Pembroke Welsh Corgi, an eleven-year-old named Andy.

Spooked by New Year's Eve fireworks, Andy had bolted while Mike and his wife, Jordina, visited friends in Westport with their three other dogs. The Ghiggeris initially assumed Andy hadn't gone far.

"He's kind of a lazy Corgi," Mike told me over lunch at Whole Foods. "He doesn't like to run very much. And he's pretty stubborn, so it wasn't that unusual when he didn't come right when we called him. We put out some treats and figured he would be there in the morning."

He wasn't.

In the weeks since Andy's disappearance, the Ghiggeris had become local celebrities of sorts for the relentlessness with which they searched for their lost pet. They'd plastered the town with posters, started a Facebook page, appeared on the local news, put out Amber Alerts, and generally acted in ways one would expect when searching for a missing child.

Mike looked exhausted on the day of my visit. He and Jordina had barely been home to Massachusetts, where they lived. And if searching for Andy wasn't stressful enough, a week before my arrival they'd had to put down their nine-year-old German Shepherd.

"She was blind and literally couldn't stand," Mike told me later as we drove around Westport in his truck, with Casey riding in the backseat. It was a gloomy, overcast day, and I could see Andy Missing signs on just about every telephone pole we passed.

Mike took us to see two of the traps they'd placed around town, both near where there had been recent Andy sightings. (The traps were filled with tasty food, including roasted chicken.) They would get sightings every few days, Mike told me, including a few miles away in Norwalk, where several people swore they saw Andy running through a residential area.

When Mike posted about the Norwalk sightings on the "Bring Andy Home" Facebook page, thirty volunteers descended on the area. "One lady we'd never met printed out two thousand flyers for us, half of them in Spanish because there are many Spanish speakers in Norwalk," he said. "People call us, crying, asking how they can help. One woman spends five hours each day looking for Andy. All for a dog they've never met."

The Ghiggeris did end up catching a dog in one of their Norwalk traps, but it wasn't Andy. It was a Sheltie mix named Lana, who'd been missing for two years. Mike thinks all the Norwalk sightings were probably of Lana.

"We've had some sightings in Westport that I believe were Andy, but the eyes see what they want to see," he told me. "And people are well-meaning, but now they're so personally invested in the search. I'm embarrassed, because a lot of people have seen people walking their Corgi around town, and people come up to them and say, 'Hey, is that *your* Corgi?'"

As we drove around Westport, I couldn't help looking down every side street, and in every backyard. When I caught a glimpse of medium-sized tan dog, I practically screamed for Mike to stop the truck. But upon closer inspection, the dog turned out to be a small, Shepherd-looking mutt.

"You get to the point where you think you see Andy everywhere," he said.

Mike estimated that he and Jordina had spent thousands of dollars so far looking for their dog. And the tab would have been higher, but many of the more expensive search items (including a dozen motion-activated night-vision cameras) were donated. The cameras — which the Ghiggeris placed near the traps, and near sightings — had sounded like a good idea, but they'd proven more tantalizing than useful. They ran on a two-second delay, meaning the Ghiggeris saw a lot of hind legs.

"We call it the *ass shot,*" he said with a chuckle. "It could be a Corgi, or it could be a raccoon."

In one of the funnier moments captured by a camera, a raccoon and a skunk duked it out over a rotisserie chicken hanging from a string.

At one point during our time together in Westport, Mike left me alone in his truck for a few minutes with Casey.

As I sat there, playing with Casey's ears, I remembered the day I thought I'd lost him. Only a few months before embarking on my journey, I'd walked Casey to one of my favorite lunch places in Jamaica Plain and tied his leash to a placard on the sidewalk. When I checked on him two minutes later, he was gone.

I have a fear of acting like a crazy person

in public, so I probably seemed reasonably calm as I stopped passersby to ask if they'd seen a dog running loose. But, beneath the surface, my heart was hammering against my rib cage, desperate to be freed from my body. If a 2011 morning show segment was to be believed, the practice of dognapping was on the rise. Hadn't they called it an *epidemic*? It all made terrifying sense: Casey was in the greedy, dirty hands of dognappers.

I started sprinting up and down the street, screaming Casey's name like a crazy person. How had no one seen anything? Why were they continuing on with their day as if nothing had happened? WHERE WAS MY DOG?

As I braced myself for the tears I knew were coming, I saw a man and a woman walking toward me. And there, trotting along at their side, was Casey. I ran toward them, screaming thanks as I approached, but as I reached for my dog the man tugged slightly at the leash to keep Casey close.

"You know, you really shouldn't attach your dog to an unsecured sign," he said, practically spitting the words at me. "It's not safe. Not safe at all."

He went on to explain that a gust of wind had knocked over the placard, startling Casey and causing him to run down the sidewalk in a panic. The placard, which listed the restaurant's specials, had rattled along behind him. The man and woman had followed Casey, and when he'd finally stopped running after two blocks, they'd untangled his leash from the placard, comforted him, and walked him back to the restaurant.

"I can't thank you enough," I told them, prying the leash from the man's fingers.

As I sat in Mike's truck with Casey, I wondered what I would have done had Casey been lost the day of the placard incident. Would I have gone to the lengths the Ghiggeris had? How many dog lovers would? Though Americans increasingly think of their dogs as family members (as "kids," or "furbabies"), people who spend an inordinate amount of time and resources searching for a lost dog — or who spend months mourning a dead dog — are often viewed with suspicion. As much as we like to talk about dogs being family members, they're still usually *second-class* family members.

When I'd told several friends (two who have dogs, and two who don't) about the Ghiggeris over dinner before embarking on my trip, three of the four expressed some degree of judgment toward the couple. They didn't quite roll their eyes, but I could tell they wanted to. One friend mentioned Andy's advanced age.

"I suppose I could understand spending thousands of dollars and putting your life on hold for a month if the dog's young and has his life ahead of him," he said. "But for an eleven-year-old that's probably near death anyway?"

Even my friend who admired the Ghiggeris for their perseverance didn't think he could match it. "I love my dog — I adore my dog," he told us. "If he got away, I'd be devastated, and I'd look for him, call local shelters, put up signs. But I probably wouldn't go on the news and put cameras around my neighborhood to find him. I don't know. Does that make me a bad person?"

When Mike returned to the truck, I asked him if he and Jordina had received any criticism for their extensive search. "Oh, yeah," he said. "Some people say, 'Oh, it's just a dog.' But I don't think they really understand what it's like to truly love a dog. To me, it's not any different than loving a kid."

The lost pet industry certainly understands. In the past ten years, there's been an explosion of products and services designed to keep your dog from getting lost — and to find her if she does. There are collars with GPS tracking devices, professional tracking dogs trained to hunt down your lost pooch (the Ghiggeris used several), and perhaps the most obvious conflation of pets and children: AMBER Pet Alerts.

Modeled after the highly successful alerts for missing kids, a handful of web-based companies promise to help you find Toto. (One is even called FindToto.com.) Some companies, including one the Ghiggeris used, send out thousands of telephone calls to homes in your neighborhood. Others rely on a nationwide network of animal lovers to search for a lost dog — and to spread the word of a missing dog — in their area.

Thanks to their television appearances, Facebook page, and thousands of Missing Dog signs, the Ghiggeris had managed to build a remarkable network in their own right. "There's no way we'll find him without the help we're getting," Mike told me as he drove us back to the RV. "It's been unbelievable. And I think the best way to thank everyone for everything is just by finding Andy."

The odds were in their favor. According to a 2012 study by the ASPCA, 93 percent of dogs reported missing are eventually found. I asked Mike how long he and Jordina would keep searching.

"I honestly don't know," he said. He seemed to be considering the question for the first time. "We want to try everything we can to find Andy, but there's the matter of our jobs, our lives. We need to be smart about this, to use our resources wisely."

As Casey and I climbed back into the motorhome to head toward New York City, where I would meet some of the craziest dog people of my journey, I did something I'd never done before: I said a prayer for a dog. Maybe, by the end of my trip, Andy would be safely home.

2.
In Which Dog-Loving Humans (Okay, New Yorkers) Behave Badly

The closest RV park to Manhattan is a gravel parking lot in Jersey City.

It's called the Liberty Harbor RV Park, and what it lacks in *park* like qualities it makes up for in "location, location, location," as one reviewer put it on RVPark Reviews.com, a site I would come to rely on. You can see the New York City skyline from your motorhome, and Midtown is a mere fifteen minutes away on the PATH train.

I pulled into Liberty Harbor on a Sunday night. It was cold, I had a migraine, and Casey needed to pee. "There are plenty of spots — I could park anywhere," I suggested to the young security guard who struggled momentarily to locate my reservation. But he was not easily persuaded, and I was relieved when he found my name on his list and escorted me to my spot near the bathrooms.

This was my first night in an RV park, and I had no idea what to do. The Chalet had a number of compartments on the outside, one of which apparently housed the thick electrical cord that I needed to plug into the thirty-amp hookup. Or was it the fifty-amp hookup? Thankfully, the guard took pity on me. He knew exactly where my electrical cord was, and that I needed to plug it in the thirty-amp socket. The moment he did, the RV's microwave flickered on.

I took my migraine medicine, led Casey for a quick walk around the parking lot, and then called my friend Jason, who lived nearby in Newark. He had agreed to visit me in the motorhome and watch Casey the next two days while I attended Westminster. But he wasn't answering his phone.

I microwaved some Stouffer's lasagna and plugged in my laptop to update my Facebook page. One can hardly take a trip these days without chronicling it on Facebook, and I'd created a special page for the occasion. I'd intended it for family and friends, but after several dog-related websites previewed my voyage, I suddenly had three thousand fans.

They were a raucous, opinionated, dog-crazed bunch. The vast majority were women. Most had rescued their dog from a shelter or rescue group. Many had their dog (or dogs) as their Facebook profile picture. And virtually all adored Casey with a fervor normally reserved for rock stars or celebrity fitness instructors.

"I LOVE you Casey!!!!!!" they would write on my wall, as if Casey spent his downtime as so many humans do — trolling social media for affirmation.

Many of my fans (or, more precisely, many of *Casey*'s fans) had long fantasized about dropping everything and hitting the road with their dogs. "Livin' the dream of so many," one poster wrote, echoing what I would hear over and over again during my months on the road. Some of Casey's fans had already lived the dream, and they emailed me daily with suggestions for my itinerary.

"No cross-country trip with your dog is complete without stopping in [insert name of town, dog park, beach, etc.]," they insisted.

In the first few days of my voyage, I'd posted a picture of Casey and Wanita on Amy Hempel's couch, as well as a video of Casey inside Amanda's studio. I'd also whined about spending $175 to fill up the RV's gas tank every four hundred miles, which had prompted unexpected offers of cash.

"My dog said he'd be more than happy to help you and Casey defray some of those outrageous gas prices," one woman wrote on the trip's Facebook wall. "Does anyone else think that would be a great way to take a little part in this trip we all wish we were on? Woof if you do!"

In the end, I decided not to pillage the bank accounts of my Facebook followers. But, though I didn't realize it that night in Jersey City, I would come to rely on Casey's online admirers for a different kind of sustenance. Throughout my travels, these strangers would seem instinctively to know what I needed — whether it was a virtual

motherly hug one lonely night in a ghostly
RV park in Texas, or a kick in the ass to
help me face an interminable drive through
Wyoming. More importantly, perhaps, they
would model how to love my dog in a manner that had somehow eluded me — uncon-
ditionally.

If only my real friends had been as helpful. At about eleven that night, it occurred
to me that Jason wouldn't be calling me
back. And I was pretty sure I knew why:
he'd likely relapsed on drugs. He was nearly
a year sober, but I'd lived through one of
his previous relapses and knew the signs. If
he's hours late and not returning phone
calls, he's probably holed up somewhere
getting high.

I swung open the Chalet's door, jumped
out, and kicked one of the RV's front tires.
This turned out to be a stupid idea, as the
migraine that had mostly dissipated returned. I turned off the motorhome's interior lights, collapsed on my bed, and began
calling friends in New York City. I reached
one, James, who doesn't particularly like
dogs but who took pity on me.

"Come by tomorrow," he said. "You guys
can stay at my place."

Before heading to the city the next morn-

97

ing, Casey and I went searching for a patch of grass. We walked along the Jersey City waterfront, past sleek boats and high-rise buildings with well-kept courtyards. Just beyond a memorial to the Korean War veterans of Hudson County, we came upon a large grassy area shaped like a shotgun, with the barrel aimed at Battery Park City in Manhattan.

There were a handful of dogs running about when we got there. I struck up a conversation with a brown-haired French woman named Julie, who was pushing her baby stroller with one hand and wielding a blue tennis ball launcher in the other.

"I'm this awful stereotype of a Jersey City mom," she told me with a laugh as her year-old Jack Russell, Maggie, zigzagged around us. "I've got the stroller, the dog, the ball launcher. You don't have to pretend to find me interesting."

But I did find Julie interesting, and not only because she gave me the chance to speak French. Before moving to the United States with her husband, Julie had what she called a traditionally European perspective on dog ownership.

"People love their dogs in France, but a dog is still considered a *dog,*" she said. "It's not usually seen as a child, like so many dogs are here."

In France, Julie had never heard of a ball launcher, let alone a doggie day spa, pet health insurance, or a raw dog food diet. In fact, she was so surprised by how Americans treat their pets that she wrote an article about American dog culture for a French magazine.

"Many French people have no idea how spoiled dogs are here," she told me.

It hadn't taken long for the spoiling to rub off on Julie. Soon after bringing Maggie home, Julie began calling the dog her baby. "Now I always say that Maggie was our first baby, and that our actual baby" — she gazed at the infant bundled up in the stroller — "is our second baby."

Julie takes both her babies to this park three times every day for an hour, even if it's "raining horizontal." All that time outdoors has the added benefit of helping Maggie lose one pound, which would allow the dog to travel with Julie and her husband in the coach cabin on an upcoming flight to France.

"She's fifteen pounds right now, but we need to get her down to the fourteen-pound weight limit so they won't put her down with the luggage," Julie explained.

"How's it going so far?"

"Not well," she said. "One pound doesn't sound like a lot . . ."

"But on a little dog . . ."

"Exactly," she said.

"Have you considered doggie fat camp?" I joked, referring to the increasing number of camps nationwide that promise to whip your dog into shape. More than half of all American dogs are considered overweight or obese. The more overweight you are, the more likely that your dog will be overweight, too.

"That's another thing that would never happen in France!" she insisted. "French dogs aren't fat."

"Maybe they're just like French people in that way?" I suggested.

It was fun to stereotype, but we were wrong. Though not as big (pardon the pun) a problem as in the United States, 39 percent of French dogs were found to be overweight or obese in a study published in *The Journal of Nutrition.*

Julie launched the tennis ball toward some big rocks at the far end of the park. Maggie and Casey both dashed off after it, but Casey gave up after a few yards when he realized Maggie would get there first. I excused myself for a moment and led Casey to the edge of the Hudson, where I hoped to take a photograph of him with the New York skyline over his shoulder. I positioned him on a rock, told him to stay, and snapped pictures with my iPhone. Then I sat next to him and tried to meditate. I'd promised my mom that I'd carve out a few minutes of quiet time each day for myself on the road, but it hadn't happened yet.

Meditating on rocks with a Lab next to the Hudson River is not easy. I tried to focus on my breath, but Casey kept barking at me to throw him the ball. When we're "on a walk," Casey considers any lollygagging unacceptable; he's never been very good at entertaining himself. I eventually gave up and returned to Julie, who was still launching tennis balls for Maggie.

"She'll run like this all day if I let her," she said.

I wished Julie luck and returned to Liberty Harbor, where I called car services until I found one willing to transport a dog. Getting a big dog into Manhattan without a car poses logistical challenges. Unless he's a service animal, he's not allowed on the ferry or the subway. And good luck hailing a cab. Nothing displeases most New York City taxi drivers more than chauffeuring a dog around.

An African American friend of mine once

told me that he loves competing for a taxi with a dog-wielding white person. "It's the only time a cabbie will choose me," he said.

I had some time to kill before meeting James at his apartment, so I took Casey for a quick stroll through the East Village, a neighborhood overrun with dogs — mostly mutts. We stopped at a dog-friendly coffee shop, the Bean, where Casey promptly knocked over a woman's iced coffee with his tail. The liquid barely missed the woman's MacBook Pro, splattering instead on the floor near my tennis shoes. I rushed to the counter in search of paper towels, but by the time I returned Casey had lapped most of it up.

"Uh-oh," I said, as a Bean employee walked briskly toward us.

To my surprise, he came bearing gifts. "I thought this big guy might want a few doggie treats," he said in a baby voice, bending down to deliver the goods into Casey's eager mouth.

"Geeeeeennnnnnnnntle," I reminded Casey, who can occasionally bite the hand that feeds him.

I led Casey over to a couch in the corner. He pulled at the leash, his nose pointing us toward a nearby woman's chocolate croissant. "Casey, down," I said forcefully. He dropped his head, sighed loudly, and stretched out on the floor at my feet.

I took the opportunity to eat a bagel and read a few pages of the book I'd brought along, *New York's Poop Scoop Law: Dogs, the Dirt, and Due Process.* Written by long-time New Yorker Michael Brandow, the book is far more interesting than its title suggests and tells the fascinating story of how dogs came to be one of New York City's most contentious political issues.

Back in the 1970s, before picking up after your pooch was an expectation of canine ownership, the city's sidewalks and parks had been littered with dog doo. In a 1975 letter to *The New York Times,* a grumpy New Yorker had lamented the sad, soiled state of Central Park: "It has become a dumping ground for defecating dogs. . . . The crisp green grass, the smell of blossoming cherry trees, the playing fields are all being taken from us by the hordes of dogs that romp freely. . . . Can nothing be done? Are we all to be slaves to people who find pleasure in keeping Great Danes in three-room apartments?"

But when New York City mayor Ed Koch proposed the country's first poop-scoop law in 1978, dog owners were aghast — and promised to revolt. "Like the Jews of Nazi Germany," the head of New York's Dog Owners Guild at the time complained, "we citizens, including the old and infirm, are being humiliated by being forced to pick up excrement from the gutter."

Many dog owners saw the proposed law as impractical (Koch made clear that dog owners were not to dispose of poop in public waste receptacles), and as a slippery slope with the ultimate goal being the eradication of dogs from Manhattan. Their paranoia wasn't entirely misplaced. Dogs and their owners had been unfairly blamed for much of what ailed New York City in the 1970s and 1980s. Dog bashing at the time was a favorite pastime of many New Yorkers, best expressed by writer Fran Lebowitz in perhaps the greatest takedown of dogs ever published.

"Pets should be disallowed by law," she wrote in 1981. "Especially dogs. Especially in New York City. I have not infrequently verbalized this sentiment in what now passes for polite society, and have invariably been the recipient of the information that even if dogs should be withheld from the frivolous, there would still be the blind and the pathologically lonely to think of. I am not totally devoid of compassion, and after much thought I believe that I have hit upon the perfect solution to this problem — let the lonely lead the blind."

But dog issues weren't just dividing New York. As dog ownership exploded across the country in the 1970s, the National League of Cities listed dog and other animal concerns as the primary issue about which local public officials received complaints.

In New York City, dog runs and parks had been proposed as early as the 1960s as a solution to "both the grass shortage and the poop surplus," as Brandow put it. (One landlord even built a rooftop dog run on his building and suggested that other landlords do the same.) But the idea that anyone should provide extra accommodations for dogs or their owners struck some as preposterous.

Many New Yorkers "were not about to hand over parcels of public land, however miniscule, permanently to those filthy, marauding beasts that had, in their opinions plagued the city." People worried that the runs would be breeding grounds of filth and disease and would invariably pollute surrounding areas. Underlying much of their anger was a belief that if a person wanted a dog that badly, then they should live somewhere *with a yard*. Even some dog owners disliked the idea of dog runs, which they saw as a kind of forced segregation. Others worried about whether their dogs would be safe there.

"Critics imagined horrible scenes of loose predators, the large animals consuming the smaller ones in a single bite," Brandow wrote.

But though there is an occasional dogfight today in New York City's dog runs, one could argue that dogs get along better there than their human counterparts. I got to see that for myself a year before my trip, when, in a moment of lunacy, I decided to spend a full day at the city's oldest dog run, Tompkins Square (also called First Run).

I'd wanted to experience the rituals and rhythms of an urban dog park. What I experienced instead would make a middle school recess — with its assorted cliques, cruelties, and comedies — feel like a conference on interpersonal ethics.

I'd awoken so early on the morning of our dog park adventure that Casey had looked at me as if I were kidding. Normally the word "walk" elicits all manner of canine gesticulation (spinning, leaping, barking), but at 5:35 A.M. on a cold weekday it elicited a strange, suspicious stare.

106

"Now?" he seemed to be thinking. "Really?"

We arrived at the park a few minutes after six (when it officially opens), but the gates were locked. Casey sniffed a fire hydrant while I tried to figure out what to do. About half a block down 10th Street, I spotted a woman and her large dog slipping through what I presumed to be a hole in the park's fence.

"We're in!" I told Casey, who pranced along next to me, game for this early-morning experiment.

The dog run is at the northern end of the ten-acre park. It's enclosed within wrought iron fences and is divided into two sections — a larger one for big dogs, and a smaller one for little dogs. A gigantic wooden dog bone sculpture attached to a three-foot-high fence divides the sections.

The woman I'd seen slip into the park was the only person in the run when Casey and I arrived. She sat on a bench in the large dog area, wearing a down jacket and clutching a cup of coffee. I sat down next to her and introduced myself. Her name was Robyn, and her dog's name was Charlie. A big, friendly black and gray Siberian Husky, he'd been a gift from her ex-husband.

"We'd separated after thirty years together," she told me, "and he gave Charlie to me hoping that I would get back with him. Basically, he tried to bribe me with a dog!"

"Did it work?" I asked.

"Well, *yes and no.* I didn't take him back at first. But, four years later, we did get back together."

"You made him sweat it out."

"Well, you know . . ." she said with a guilty smile. "Once I had Charlie I didn't even really feel like I *needed* a husband. You know what I mean? Dogs can be a lot more fun than husbands. And Charlie is the most amazing gift in my life. Dogs are the most amazing gift! But you know that. I love your dog. She's beautiful."

It wasn't the first time someone mistook Casey for female. I'm not sure if it's his name, his soft white fur, or his blond eyelashes, but my dog could win a doggie androgyny contest.

I asked Robyn if she ever worried about owning a Siberian Husky in Manhattan. After all, she'd told me she'd lived in the same small, rent-controlled East Village apartment since she was a child. Was that really the ideal place for a dog originally bred to pull heavy loads across eastern Siberia?

"Are you kidding?" she blurted out. "Charlie has the best life ever. I'm out here three or four times a day with him, in between my jobs as a housekeeper. He comes out here every morning at six for an hour and a half. I'm part of the morning crew."

"The morning crew?" I asked.

"There are about six or seven of us who come here each morning when the park opens," she explained. "We've become great friends over the years through our dogs. I mean, what other people besides your family and your co-workers do you get to see for an hour every day?"

Just then, another member of the crew arrived with two dogs in tow. "This is Sean," Robyn said, introducing us. "He's a dog walker and has been coming here longer than just about anybody else."

"Yeah, I have no real life to speak of," he said, eyeing me with suspicion. "But enough about me. Who the hell are you?"

"This guy's writing a book about dogs," Robyn explained. "He's spending all day at the run."

"All day at the run?" Sean guffawed. "Dude, you're nuts. And your poor dog! He's going to hate you by the end of the day." Sean pulled out his smartphone.

109

"What's your name? I'm going to Google you."

"You're going to Google me?"

"Yeah, gotta see if you're a real writer. Can't be having strange people hanging out all day at the run pretending to write books about dogs."

We heard the clacking of the gate at the entrance to the run. Linda, a woman in her early forties, entered with her retriever mix, a cup of coffee, and that morning's *New York Times*.

"This dude's writing a book," Sean told her as she plopped herself down next to me.

"Congratulations," she said.

"No, about *this dog run*," he clarified.

"Oh, well, there are plenty of crazies here. You'll have lots to write about."

Sean laughed. "There's this saying — to own a dog in Manhattan, you have to be crazy," he told me.

"Or rich," Linda said. She looked at me, then Casey. "Is this your yellow Lab?"

Casey stuck his nose in her lap, and she fawned over him for a few seconds before her dog started humping him. "He likes Casey!" Sean said. As if on cue, Casey starting humping Charlie.

I asked the morning crew if any celebrities were regulars at the run. Sean proudly ticked off a dozen names, including Molly Ringwald, Parker Posey, and Josh Lucas. Linda rolled her eyes. "Sean, you're such a star-fucker."

"There's a lot more," he went on, undeterred. "Elijah Wood comes here sometimes. You know, the *hobbit*. He has a girlfriend who lives nearby and who has three small dogs. He's as short as she is! I swear he's like four foot nine. Oh, and Matt Dillon" — he pointed to a building on the corner — "lives right there."

"He doesn't have a dog, though," Linda said. "He just likes the attention."

Sean rolled his eyes. "A dog park is really a great place to meet people," he told me. "We've actually had a few marriages happen between people who first met each other here."

"I try not to date where my dog shits," Linda said.

Linda used to live uptown and was a regular at the dog run in Madison Square Park, which she recalled as considerably less crazy than this one. "Down here, there's so much drama," she said. "The only time I can tolerate this place is early in the morning. The lunatics come out later."

"So you guys don't consider yourselves lunatics?" I asked with a smile.

"There's normal New York crazy, which is what we are," Linda explained. "Then there's batshit New York crazy, the people who have nothing better to do than start drama in the dog park. And don't even get me started on the people over on the little dog side. They're an entirely different kind of crazy."

Later that morning, the cutest puppy I'd ever seen — a four-month-old Golden Retriever named Bean who looked like the dog in the Snuggies television commercials — bounded into First Run.

There were about fifteen people there when Bean arrived, and for a good five minutes nearly everyone stopped what they were doing to take turns petting her.

"Having a cute dog in New York City makes living here so much cooler," Bean's owner, a friendly photographer in his thirties, told me. "I'll be walking her down the street, and people will just grin when they see her. People stop and pet her and talk to me. The biggest, meanest-looking guys are like, 'Oh, it's a PUPPY!' "

(Had the photographer been in the market for a date, Bean would have proven even more useful. In a French study, women were significantly more likely to give their phone number to a man who asked for it when he was accompanied by a dog. Findings from another study — by researchers at the University of California, Santa Barbara — suggest that women are especially drawn to younger dogs. The study found that a Golden Retriever's ability to entice strangers to pet it declined as it aged from ten to thirty-three weeks, with women growing markedly less responsive as the dog grew older. "The results suggest a human, and especially a female, preference for canine juvenescence," the researchers wrote.)

While Bean played with a stick, I meandered over to a young dog walker named Ryan. He was at the run with two dogs, including a friendly Pit Bull named Bullet. A regular at the run, Ryan told me he came here three times each day with a total of ten dogs. He likes holding in-depth conversations with them. He calls them "knuckleheads," asks them how they're doing, and philosophizes about their inner thoughts.

Growing up, Ryan never thought he'd walk dogs for a living. In fact, he stumbled into the job after several years of bartending and an unsuccessful one-day stint as an office temp. "My boss at the office was like, 'Wow, you really suck at this,' " Ryan recalled. "But he had a friend who was looking to hire another dog walker for his dog walking company. So he put me in touch, and I started my dog walking career the next day. First job I've ever had with health insurance!"

Though he loves dogs, Ryan doesn't have one of his own. "I spend all day out walking dogs," he explained, "so if I had one at home, I'd have to hire a dog walker for him. That just seems silly. I have two cats. No need to hire a cat walker."

When Ryan left the run, I spent some time throwing a tennis ball for Casey. It was colder than I expected, and I'd under-dressed by wearing only a sweater. A few minutes later, I noticed a pale, dark-haired young woman sitting on a bench near the run's entrance. She sat staring straight ahead, her small hands folded in her lap. Though her eyes were open, she appeared to be meditating.

I sat down next to her. As I did she smiled warmly and rotated her body — almost ro-botically — toward me.

"Hello," she said.

"How are you doing?" I asked.

"I just got out of a mental hospital, actu-ally," she said as her dog sniffed Casey's rear end.

I didn't know what to say to that. I

scanned the dog park for a hidden camera.

"I've had really severe OCD for four years," she went on to say. "I'm an actress, and it's been so bad I haven't been able to work for the last year. But I'm much better now that I'm off my medications. I had a sort of intuitive dream that God told me to get off the medicine. It was all making me feel even more bonkers. It gave me seizures, and I would only talk in perfect Shakespearean rhyme for like two or three days straight, which made everyone in my life uncomfortable. They worried about me, you know. I also talked that way in the hospital. I would have everyone in stitches because I'd talk about the nurses, the doctors, everything, in *perfect rhyme*! I would rhyme about how the medicine we were all on wasn't letting us feel our real feelings, and soon I was triggering all the patients in there to laugh and cry and have breakdowns — but the good kind of breakdowns, you know? I believe in God — I don't know if you do — but I really do."

Just then a helicopter flew over the park. Several dogs in the run stopped what they were doing and looked up in the sky toward the loud, strange noise.

"Since I've been out of the hospital," she continued when the chopper was gone and it was possible to hear each other again, "I try to come to the run as often as I can. It helps me get back to my center, and it's one of the highlights of my day. I missed my dog so much in the hospital. I just have to wink at her, and she knows what I want to do. She's very in tune with me. Very in tune! When I was on a plane and having seizures because I was in withdrawal from my medicines, she started having seizures, too. It's like she was feeling everything I was feeling."

She went on like this for a few minutes, stopping occasionally for air. Then she changed the subject.

"Do you have a girlfriend?" she asked.

"I'm gay, actually," I replied. I could imagine a guy offering her a similar answer even if he wasn't gay, but in my case it happened to be true.

She cocked her head slightly to the side and squinted her eyes.

"Really? You don't *seem* gay. You don't have that voice."

"We come in many voices," I assured her.

"I'm not sure that's true," she said. "But you've never been attracted to women?"

"Not in the same way heterosexual men are attracted to women."

"Maybe if you met the right woman," she

insisted. "Men and women really just fit together more naturally. You just have to look for a restored version of Eve. There are a few women, including myself, who are restored to their full womanhood."

At that moment, Sean entered the run with two new dogs. (He has different shifts for the dogs he walks.) I never thought I'd be so happy to see him. "I need to go over there and ask that man a question," I lied. "I hope you'll excuse me."

The woman smiled politely. "You'll find the right woman," she said. "I'll pray for you."

Sean seemed equally happy to see me. When I was within shouting distance, he smiled broadly and yelled, "Dude! I have good news, and I have bad news!"

I asked for the good news first. "I looked you up when I got home," he said. "You checked out — you're apparently a real writer."

"And the bad news?"

"Your last book's available on Amazon for one cent," he said with a chuckle. "Hopefully this one will fare better."

I told Sean about the woman on the bench. "She was either crazy, or the best actress in town," I said. "And I think she was hitting on me."

"Oh, you'll get hit on by all kinds of crazy people here," he said. "Best to just get used to it."

I suggested we check out the little dog area. We stopped short of going in — Casey's too big, as were Sean's dogs — but we stood next to the fence that separated the runs and watched a Boston Terrier put a little white fluffy dog into what appeared to be a headlock.

"Brutus, no headlocks!" screamed the dog's owner, a man in a New York Jets cap.

Sean then introduced me to a park regular who was there with his eleven-month-old terrier. For many years, the guy had come to the run with a 120-pound Mastiff. Then he went a few years without a pet.

"But my children were becoming petless freaks," he told me. "And I couldn't have that."

"Petless freaks?" I asked.

"Yeah, there are all kinds of studies that show that kids with dogs are better off," he said.

He was right. Researchers have found that children deeply value their relationships with dogs, and that those who grow up with them are healthier (they're less likely to suffer from asthma, for example) and have higher degrees of empathy and self-esteem.

Psychologists at Oregon State University found that preschool children who were given a puppy to look after not only gained confidence and empathy — they also turned out to be more popular than puppyless kids. Looking after the puppy "made the children more cooperative and sharing," researcher Dr. Sue Doescher discovered. "Having a pet improves children's role-taking skills because they have to put themselves in the pet's position and try to feel how the pet feels. And that transfers to how other kids feel."

But not everyone who grows up with a dog turns out to be popular, cooperative, or empathetic. For proof of that, I just needed to dig deeper into the lives of many of the regulars at this run — many of whom grew up with dogs. Sean described this place as a kind of "war zone," and over the next few hours, I would come to understand what he meant.

"I'm from a small town in Texas, so I'm used to dealing with idiots and bullies," said Frank, a handsome man in his mid-thirties. We were sitting on a bench in First Run later that afternoon while his dog chased a mutt who took refuge under a picnic table. "But what I've seen in here is really beyond the pale. A lot of these people are nice until you piss them off, and then they will get personal and try to upend your life."

"What kinds of things do they say?" I asked.

"Some people said I'd been to jail," Frank told me, shaking his head. "It's out of control. A lot of regulars have stopped coming here because they don't want to deal with the drama. This run has a lot of bitter people. And I'm probably one of them. One guy got so upset by something another dog owner said that he pushed him right over the fence."

"Really? Over the fence?"

"Yup. Right over the fence."

I chuckled. "And people say *dogs* need to be leashed."

"People are much worse than the dogs," Frank insisted. "It's funny to see all the dogs of rivals playing together. We'll just kind of shoot dirty looks to each other from across the park while our dogs have fun. People will mutter under their breath, 'Anna, get away from him!,' but the dogs don't know any better."

Frank turned to face a friend of his, a beautiful young woman who sat cross-legged next to us on the bench. "Did Sean and company ever warn you about me, or did they just give you dirty looks when you talked to me?" Frank asked the woman, who looked like she belonged on a Times Square billboard.

"Just a few looks," she replied, seeming bored by the conversation.

Frank turned back toward me. "They're definitely not happy we're talking," he said, referring to Sean and the run's elected manager, Garrett, who were huddled together on the other side of the park. (Frank was wrong about that, though. Sean had pointed him out and suggested I speak to him in order to get a complete picture of the run's warring factions.)

"But how did all this drama start?" I asked. "What's it all about?"

"What's it *not* about?" Frank said, sounding both resigned and deeply philosophical.

In a 2003 *New York Times* piece headlined "Straining at the Leash," writer Caroline Campion tried to unpack the breadth of contentious issues that have faced the Tompkins Square dog run. She found that "all manner of ethnic, racial, economic and class-related conflicts have played themselves out at the run. Gentrification, ownership of public space, animal rights, self-expression, politics — all these issues have reverberated in the dogs' play space." A park regular told her, "You could propose putting the Fountain of Youth in the middle of Tompkins Square Park, and people would get upset. There would be controversy."

In the last decade, park regulars have clashed over changing the run's surface from wood chips to a sandy, crushed-rock substance; they've argued about whether small dogs should get their own section; they've revolted over an alleged lack of transparency about the park's finances; and they've armed themselves with knives in response to several Pit Bull attacks in the run.

Underscoring much of the dysfunction are class and racial tensions. In 2003, someone crossed out a sign that read "This is a community park" and scribbled, "This is a Yuppie catering service." One park regular had a dog that would attack smaller dogs. "When confronted," Campion wrote, "the woman complained that the problem was the new people . . . whose dogs deserved getting roughed up because they wore sweaters."

And then there are those who wish First Run wasn't there at all. "Homeless people were moved out to make room for the dogs, dogs that have been enslaved for domesticity," a homeless man said. "They should be in the countryside. Not a pleasure for rich people, the rich homosexuals and freaks of society."

Many dog park squabbles across the country have an undercurrent of class warfare. In Avon, Connecticut, a fierce battle erupted between longtime residents and affluent newcomers who'd moved into recently built homes. The latter group often commuted into Hartford, owned dogs, and wanted to run them off-leash in a public park. Many residents objected. Both groups formed action committees — "Save Our Park" vs. "Trails for Tails" — and claimed to be fighting over dogs and public space, though one could argue they were actually fighting over changing demographics and the soul of a town.

When I approached Garrett later that afternoon to talk about the run's various controversies and personality conflicts, he initially told me he wasn't in the mood — his twelve-year-old Rhodesian Ridgeback had died a few days earlier.

"I'm here physically, but I'm not really here," he said. "I'm still in a daze."

Later, though, Garrett opened up. "Dog parks are a lot like post-Apocalyptic culture," he said as we sat on a picnic table. It was mid-afternoon, a slow time at the park.

Casey was exhausted; he'd just finished a round of humping and was now sprawled out on the ground next to our table. "We're all in this little piece of land, having a weird kind of power struggle and attempting to get along," Garrett continued. "You're forced to interact with people you normally wouldn't. And every nut job in the world comes to a public park. Thank God for our dogs, or this place would be unbearable."

As we talked, a mustached man in a red scarf leaned his chest over the run's fence and called out to any dog that might listen. Casey lifted his head and looked toward the guy, who kept saying, "Who wants a pet? Who wants a pet?" Likely sensing the possibility of food, Casey stood up on all fours and shuffled toward the man, who smiled at my approaching dog and clasped his hands on the fence in excitement. Casey accepted a few pets before realizing the man didn't have the goods.

"Where are you going?" the man lamented as Casey walked back toward us.

"Do you think people sometimes forget about everything that's good about this kind of park?" I asked Garrett. "It seems like I've heard a lot of complaining today, and not a lot of appreciation." I pointed toward the man in the scarf. "All day, I've seen people walking by and stopping just to watch the dogs play. It seems like First Run is so important to this community, whether you have dogs or not."

"It is," Garrett said. "Even with all the drama, people come here because it's a community; it's home. When 9/11 happened, everyone came rushing to the dog park. People felt lost, and we all wanted to be with others who would understand. And in this city where people don't always feel connected, where so many people pay $2,500 a month to live in a shoebox, it's telling that so many of us rushed straight to the run. It was the natural place to come."

"Did everyone forget about their grudges for a few days?"

"Yes, all of that didn't feel important anymore. But we're human, and that feeling fades. And then people complain, especially if anything changes. The park is a lot cleaner now than it used to be, but some people — especially in New York City — get attached to their urban blight zones."

Though I'd intended to stay at First Run until sunset, after ten hours I desperately needed a change of scenery. Before leaving the park, though, I left Casey in the big-dog section and walked over to hang out with the little dogs. It was quiet over there, except for an occasional small-dog yap. I struck up a conversation with three women in their twenties and told them all about the trouble over on the big-dog side.

"We have our drama here, too," one assured me, "but we're way more passive-aggressive about it."

"We have our crazies," another said. "One woman wanted her dog to give birth in the run. We were like, 'Um, no, they have vets for that!' "

I looked back at Casey, who stared sadly at me through the fence.

"You can totally bring him in here if you want," one of the young women told me. "The people who would start drama about it aren't here right now."

"I actually need to get out of here," I told them. "I've been at the run since six this morning."

They looked at me like I had three heads. "That's the most ridiculous thing I've ever heard," one said.

On my way out of First Run with Casey, I bumped into Sean, who was walking back into the park with more dogs. He wouldn't let me leave without a fight.

"You're leaving?"

"I have to get out of here," I said.

"You got a problem with this place?"

"It's a bastion of free love and goodwill, to be sure," I said with a smile.

Sean laughed and triumphantly raised his arms toward the sky. "Hey, man. If you can survive a dog park in New York City, you can survive anywhere!"

Two miles from Tompkins Square Park, a different kind of human-canine drama was playing out at the Westminster Dog Show, the two-day all-breed dog show considered the Super Bowl of canine breeding. Westminster was first held in 1877 and is the second longest continuous sporting event after the Kentucky Derby.

After dropping off Casey at James's apartment, I took a taxi to Madison Square Garden. I made my way to the arena floor, which was covered with green carpeting and divided into six roped-off show rings for the breed competitions. Each breed winner goes on to one of seven group finals — sporting, hound, working, terrier, toy, nonsporting, and herding. The seven group winners compete for Best in Show.

A large crowd of onlookers had gathered around Ring 2, where the Xoloitzcuintli breed (Xolo for short, pronounced SHO-lo) appeared in its first Westminster since 1955. A rare, ancient breed dating back some three thousand years, early Xolos were believed to be hairless mutations of indigenous dogs. The Aztecs considered the Xolos to have healing powers, and effigies of the dogs have been found in Aztec tombs.

Partly because of this mystical history, and partly because we're suckers for strange-looking dogs, the breed had commanded much of the media's attention in the lead-up to Westminster and was even featured in the Fashion & Style section of *The New York Times.*

"Admirers of the Xolo concede that the dog is plug-ugly," the *Times* noted. "One description of this hairless canine of ancient lineage, a national treasure in its native Mexico, characterizes the Xolo as a hot water bottle with pig eyes, bat ears and a rat tail. That is being polite."

The breed comes in toy, miniature, and standard sizes, though they were all pitted against each other at Westminster. The favorite was a sleek standard-sized Xolo named Giorgio Armani, a fitting name considering it was also Fashion Week in New York City. The dog didn't disappoint. When the judge pronounced him the breed winner, he posed stoically for photographs while the rest of the Xolo handlers wondered what might have been.

"She decided to be goofy today," one handler said of her three-year-old Xolo, whose jocular temperament had failed to charm the judge. "She wasn't herself in the ring, so it threw me off."

But the woman didn't seem heartbroken, and she echoed other handlers who knew their dogs had little chance at winning. Though every dog that competes at Westminster has earned enough points in previous breed competitions, there are clear favorites. "It's a cliché, but I'm just happy to be here," she told me. "And I love this dog, so even though she didn't show well today, we'll still take her home!"

The mood was significantly less carefree around Ring 1, where Poodles pranced across the carpeted floor, their heads held high in a seemingly conscious display of their rarefied status. Though the breed's popularity has decreased since its peak in 1969, it's still the show dog of the One Percent. Not coincidentally, the breed always seems to be an early favorite for Best in Show.

This year was no exception, and the only question seemed to be which of the two marquee Poodles — a black dog named London, or a white one named Ally — would represent the breed in the group competition. There was an audible gasp, then, when the judge selected a long-shot Poodle from Canada.

"What just happened?" one shocked onlooker asked her friend.

"I think the Canadian dog won," the friend said, shaking her head.

"The *Canadian* dog?" she sneered, her face contorted in disgust.

Was the Canadian dog really better? I had no idea — the Poodles all looked the same to me. Even for those with a trained eye, show judging is highly subjective. Judges are supposed to decide which dog best represents the breed standard (a written blueprint for how a dog should look), but barely an hour goes by at Westminster without someone whispering that a particular dog won because a particular judge is friendly with a particular handler.

As you know if you've seen the film *Best in Show,* what happens in the ring isn't even the most interesting part of Westminster. The real action happens in the benching area, where handlers frantically prepare their dogs for competition while thousands of fans shuffle through armed with cameras, questions, and requests to pet the merchandise.

The benching area is not the place for the

claustrophobic or the easily irritated. No-where else on earth do so many humans and dogs convene under one roof, and to call the result a clusterfuck is to understate it. In the hour I spent back there during the first day of competition, I saw a small child run facefirst into an unsuspecting Rhodesian Ridgeback, a humorless female handler curse out her assistant, a self-described pet psychic hawk her services, and two invasions of personal space that nearly escalated into blows. I also nearly stepped on a Beagle, who was busy sniffing something under a dog crate. (Dog trainer and animal behaviorist Patricia McConnell calls the Beagle, which has an uncanny sense of smell, "the nose with paws.")

As I waded through the crowd, I hoped to bump into some Best in Show favorites, including a Pekingese named Malachy. But Malachy's owners — including Iris Love, a privileged child of the Depression who gained archaeological fame in 1970 as the excavator of the Temple of Aphrodite — wisely kept the eleven-pound dog out of sight.

The Pekingese, after all, is an easy kidnapping target. For one thing, it can barely move. For another, it can be stashed inside loose clothing. In his book *Show Dog,* Josh

Dean writes that the breed was designed to "fit inside the sleeves of a Chinese nobleman's robes in the times before central heating."

Several legends purport to explain how the Pekingese came to be. My favorite is that a lion fell in love with a butterfly, but that the pair quickly realized their size differential might be too massive a hurdle to overcome. Looking for answers, they traveled together to meet the Buddha, who graciously split the size difference and created the Pekingese. As the legend goes, the dog was kept in palaces and temples to search out and destroy demons that infiltrated them.

Today, it's difficult to imagine a flat-faced Pekingese walking around the block — let alone slaying demons. Malachy was by far the least athletic of the seven dogs that competed the next day for Best in Show, among them a Dalmatian, Irish Setter, German Shepherd, and Dachshund. While the other dogs easily navigated the show ring, Malachy moved marginally faster than a snail and needed to rest on a bag of ice when he was done.

"What do you think of the Pekingese breed?" I asked Josh Dean, whom I sat next to the following day for the finals.

"It's not so much a dog as it is a kind of companion for a *Star Wars* character," he said. "It can barely breathe, barely move, and looks like something you'd put your feet on when watching television."

Still, Malachy went on to win Best in Show. The controversial decision played perfectly into the narrative articulated by the increasing number of Americans who eschew purebreds in favor of rescues and mutts — namely, that breeders and the American Kennel Club are more concerned with aesthetics than health. In their book *Dogs,* Raymond and Lorna Coppinger argue that breeders and the AKC — which has seen registrations nosedive since the 1990s — are "producing unhealthy freaks to satisfy human whims."

Many in the media saw Malachy as a prime example of the problem. In the days after Westminster, the press would go on to dismiss Malachy as a "walking dust mop," "fuzzy bedroom slippers," and "Geraldo Rivera's mustache."

Underlying much of the media coverage was a belief that Malachy, though technically a dog, hardly represented an active and healthy pet.

The same could be said for a number of

other breeds at Westminster, including one I'd spent some of the previous two years studying — the Bulldog.

I didn't have many friends among the Bulldog breeders and handlers at Madison Square Garden. That's because three months before the show, I'd written a cover story for *The New York Times Magazine* about the breed's health problems. I'd argued that the modern Bulldog is unique for the sheer breadth of its dysfunction. Studies have shown that Bulldogs are significantly more likely than other dogs to suffer from a wide variety of ailments, including ear and eye problems, skin infections, immunological and neurological problems, and locomotor challenges.

Because of their flat faces, Bulldogs are synonymous with brachycephalic airway syndrome, which comprises a series of respiratory abnormalities affecting the throat, nose, and mouth. Dr. John Lewis, an assistant professor of dentistry and oral surgery at Ryan Veterinary Hospital of the University of Pennsylvania, said that the human equivalent to breathing the way some Bulldogs do "would be if we walked around with our mouth or nose closed and breathed through a straw."

Yet despite overwhelming evidence to the

contrary, most Bulldog breeders — including several in attendance at Westminster — flatly reject the idea that anything seriously ails the breed. A typical response came from the Bulldog Club of America, which told me that "Bulldogs today look good, have excellent temperament, and are healthier than in years past as a result of good breeding practices."

I found their denials delusional, and my *Times* story took them to task for turning a once athletic breed into a plodding, dysfunctional mess.

"Is there a bounty on your head?" Josh Dean joked near the Westminster benching area when I nearly bumped into Jay Serion, a Bulldog handler and breeder whose bulldog had won the breed competition here the previous year. I'd interviewed Serion before and after that victory, but on this day he was too busy to notice me.

I'd first become interested in Bulldogs back in 2009, shortly before attending a football game between the universities of Georgia and South Carolina in Athens, Georgia. I spent much of that game on the sideline next to the air-conditioned doghouse of Uga VII, the school's Bulldog mascot. The dog wore a red Georgia jersey and spiked red leather collar, and every once in a while he would be led onto the field to pose for pictures and model his wrinkly, smooshed Bulldog face for ESPN's cameras.

At the game I met Sonny Seiler, a lawyer and the mercurial owner of the Georgia Bulldog mascot dynasty. Sonny bore a striking resemblance to the mascots — all called Uga — he has cared for since 1956. He had a round, droopy face and wide, slumping shoulders, and his courtroom antics have often been described in words associated with Bulldogs: *Georgia Magazine* said he possessed a "barrel-chested bravura," while John Berendt wrote that Sonny "thunders and growls" in *Midnight in the Garden of Good and Evil,* in which Sonny is a character. (He defended a wealthy antiques dealer charged with the murder of a young hustler.)

Sonny wasn't in the best mood during my visit to Athens; he was tired of journalists asking him about the health of the Bulldog breed. Earlier that year, Adam Goldfarb of the Humane Society of the United States had told *The Augusta Chronicle* that Bulldogs were the "poster child for breeding gone awry." Goldfarb's quote came in response to a scathing British documentary, *Pedigree Dogs Exposed,* which highlighted

136

the health and welfare problems of purebred dogs and claimed that breeders and the Kennel Club (the British equivalent of the American Kennel Club) were in denial about the extent of the problem.

Broadcast on the BBC, *Exposed* spawned three independent reports into purebred breeding, each finding that some modern breeding practices — including inbreeding and breeding for "extreme traits," like the massive and short-faced head of the Bulldog — are detrimental to the health and welfare of dogs. All three reports called the modern Bulldog a breed in need of an intervention.

"Many would question whether the breed's quality of life is so compromised that its breeding should be banned," Dr. Nicola Rooney and Dr. David Sargan concluded in one of the reports, "Pedigree Dog Breeding in the U.K.: A Major Welfare Concern?"

Sonny dismissed any talk of changing Bulldog breeding practices, and he insisted that Uga VII was a vigorous animal who enjoyed his mascot duties. I wasn't so sure. During my visit, the dog seemed most comfortable in the back corner of his doghouse — or, better yet, outside the stadium entirely. A few minutes before halftime, Seiler's adult son, Charles, led Uga

VII off the field by a leash to a waiting golf cart. The dog hopped on, and a young woman drove us out the stadium's back service entrance, up a hill, around some bends, to an unspectacular patch of grass that doubled as his game day bathroom. When the cart came to a stop, Uga VII bounded off it and spent the next few minutes happily sniffing the grass, urinating on a tree, and defecating behind a bush.

When the dog was done, Charles ordered us all back on the cart. "All right, let's go," he said, and before I knew it, we were speeding back toward Sanford Stadium, Uga VII's droppings (Charles didn't pick up after him) a reminder to all that the world's most famous mascot was here — and that celebrity dogs, like their human counterparts, get to play by different rules.

But Uga VII's celebrity life would be short-lived. Six months later, while lounging at home, he died of heart failure. He was four years old.

When I returned to Georgia the following year to meet the school's newest mascot, Uga VIII, Sonny insisted that the eleven-month-old was "a damn good dog. He's healthy, and he has all the attributes we look for in a Georgia mascot." But by the time he turned two, Uga VIII came down with lymphoma. Two months before I embarked on my cross-country trip, he died.

Despite their health problems, Bulldogs have skyrocketed up the AKC's most popular breed list, from No. 41 in 1973 to No. 5 in 2013. James Serpell, the director of the Center for the Interaction of Animals and Society at the University of Pennsylvania, partly blames the breed's fame on a phenomenon called "anthropomorphic selection." He argues that we've bred dogs like the Bulldog (and other short-faced brachycephalic breeds, including the Pug and the French Bulldog) in ways that "facilitate the attribution of human mental states to animals."

"We have, to some extent, accentuated physical characteristics of the breed to make it look more human, although essentially more like caricatures of humans, and specifically of children," he told me. "We've bred Bulldogs for their flat face, big eyes, huge mouth in relation to head size, and huge smiling face."

Advertisers and animators have long recognized that giving an animal big eyes and a big head is a surefire way to endear it to humans. When Walt Disney created Bambi, the studio wanted the character to be an accurate depiction of a deer. But when the original Bambi sketches were deemed not cute enough, Disney shortened Bambi's muzzle and made his head and eyes bigger.

In an essay in the anthology *Thinking with Animals: New Perspectives on Anthropomorphism,* Serpell wrote that "if bulldogs were the product of genetic engineering by agri-pharmaceutical corporations, there would be protest demonstrations throughout the Western world, and rightly so. But because they have been generated by anthropomorphic selection, their handicaps are not only overlooked but even, in some quarters, applauded."

Neither Uga VII nor Uga VIII — nor the Bulldogs that Josh Dean and I watched wobble around the show ring at Westminster — looked anything like the Bulldog of the early 1800s, which were used for bull baiting, a brutal sport where a dog would attack the nose and head of a bull. Back then, Bulldogs were leaner and higher off the ground, and their muzzles longer. They also had smaller heads and fewer facial rolls.

Because of their viciousness in the ring, Bulldogs of that period were sometimes dismissed in similar ways as modern Pit Bulls. In an 1845 book, *The Dog,* a veterinarian wrote that the Bulldog was "scarcely

140

capable of any education" and "fitted for nothing but ferocity and combat." But as nineteenth-century England went purebred dog crazy ("Nobody who is anybody can afford to be followed about by a mongrel dog," one dog publication claimed at the time), the breed underwent a physical, temperamental, and public relations transformation. Famed British dog dealer Bill George, who had worked at a kennel that bred dogs for bull baiting, turned his attention to breeding and promoting the Bulldog as a pleasant household pet and stylish purebred companion.

Just how breeders like George succeeded in changing the look and temperament of the Bulldog is a debated question. Many believe that the breed was crossed with the Pug, creating a friendlier and more compact dog with a brachycephalic skull. Breeders have accentuated the Bulldog's smooshed face and compact body since then, creating a dog with a fundamentally flawed design. But breeders are either incapable of conceding those flaws — or are blind to them.

"Breeding certainly has a place in the world of dogs," Wayne Pacelle of the Humane Society of the United States told me, "but this mania about achieving what's considered a 'perfect' or desirable outward appearance, rather than focusing on the physical soundness of the animal, is one of the biggest dog welfare problems in this country."

I asked him if we should start calling irresponsible dog breeding a form of animal cruelty.

"Yes," he said. "That would be a good start."

3.
In Which Casey and I Encounter Cats, Cows, PETA, and the "Dog-Poop Lady"

After several days in Manhattan, Casey had surely forgotten about the existence of the Chalet.

He seemed understandably displeased,

then, when we returned to the Liberty Harbor RV Park in Jersey City. As we approached the RV's side entrance, Casey stopped in his tracks and looked up at me with sad, disbelieving eyes. To counter that, I started jumping up and down and pointing dramatically toward the RV, as if something life-changing awaited us inside. When Casey's tail began to flutter, I leaped in one motion up and into the motorhome and called for my dog to join me. He bought it. I gave Casey a treat and apologized for my chicanery.

The RV was the same as I'd left it, which is to say it was messy. I'd spent only a few days living in it, but there were already dirty dishes in the sink, wrinkled clothes on the bed, rogue papers on the desk. Even a miniature pet lifelike sculpture of Casey (with a natural fiber coat) prepared for me by dog-obsessed artist Lucy Maloney had fallen off the dashboard and was wedged under the driver's seat. I'd harbored grand plans of a personality change during this road trip. What better time to morph into a neat freak than during a cross-country journey in an RV? And because a motorhome is both an apartment *and* a car, by keeping it presentable I figured I'd be killing two birds with one stone.

But that delusion had shattered all over the inside of the Chalet. And the problem with a messy motorhome is more than just aesthetic. If things aren't put away and tied down, they become dangerous projectiles as you drive. Pots slide off the gas range and crash to the floor, scaring the daylights out of Casey; laptops fly off the desk, landing on Casey's head. Both had happened only minutes after departing from Provincetown the previous week. I'd been too embarrassed to tell anyone.

After tidying up (and securing) the RV, we left Jersey City and drove toward College Park, Maryland, a suburb of D.C. The four-hour drive — down the New Jersey Turnpike, and then south on I-95 — was unremarkable, so I passed the time by catching up on episodes of NPR's *This American Life.* One was titled "In Dog We Trust," but it turned out, disappointingly, to be mostly about cats.

The program got me thinking about my unexpected foray into cat ownership. Back in 2010, a friend from Franklin Park in Jamaica Plain had waved me down as I drove to the post office. She looked harried, as was her custom, but she seemed particularly discombobulated that morning. I rolled down the passenger-side window to see what the fuss was about.

"Do you want a cat?" she blurted out, leaning inside the car and staring at me with the slightly unhinged look common to those who rescue cats. "I've got a great one-year-old that needs a home."

It wasn't the first time she'd asked me if I wanted a cat. She was a one-woman cat-saving operation, using her free time to rescue them from a nearby neighborhood with a feral feline problem. She would then stash them in her office at work and try to palm them off on her friends.

I was not an easy sell. "You know what I think about cats," I told her. "They're moody. And they have claws. *Like a tiger.*"

But then she said something astonishing. "Benoit, this cat doesn't act like a cat. He acts like a dog!"

A cat that acts like a dog? I was intrigued.

"You'll love him," she continued, sensing my weakness. "Why don't you meet him? Take him home for a few days and see how it goes."

So I did. I named him Eli, and in many ways he did act like a dog. If I wanted to wrestle, he would wrestle. If I wanted to throw him in the air like a football, he would let me throw him in the air like a football. He loved greeting strangers at the door. Being sprayed with water — normally a cat's worst nightmare — hardly fazed him. He was so doglike, in fact, that one morning I took him to Franklin Park with Casey and tried to walk him on a leash. Eli didn't move, and Casey barked at me to hurry up.

"Okay, so you're not actually a dog," I conceded.

There was another important way Eli didn't act like a dog: He went to the bathroom indoors. I struggled mightily to accept this reality of indoor cat ownership. I tried several brands of cat litter, but none could compete with the noxious odors emanating from Eli's intestines.

He also rarely slept. My cat survived, it seemed, on a mere three-hour nap each day, from precisely one to four each afternoon. To be fair, Eli may have also dozed off for an hour or two at night, but I wouldn't know because he wasn't allowed in my bedroom. I'd hoped to be able to snuggle up in bed next to Casey and Eli, but Eli treated my bed as if the sheets were made of catnip. He would duck under the covers, hurling himself at my feet and generally engaging in his own Kitty Olympics. I would try my best to wait him out, but he never let up.

One day it occurred to me that Eli might

need a friend to play with, someone to keep him company and tire him out. And that's how I — a dog person — found myself with a second rescue cat. I named her Chi (pronounced Chee). She was orange, and she looked like a miniature tiger. But she was very much a cat; curious but careful, and affectionate on her own terms. She mostly ignored me at first, but she got along great with Eli and quickly warmed up to Casey.

Problem was, I had little patience for my cats' neuroses. They would sit outside my bedroom door at 5:30 each morning meowing their heads of. (My boyfriend at the time threatened to drown them in Jamaica Pond.) There was also the problem of my landlady, an elderly Greek lady who lived upstairs and who grudgingly accepted Casey but inexplicably despised cats. She was not pleased to see Eli and Chi relaxing by my living room window, staring at her as she watered her plants. She finally told me that if I didn't get rid of them, she wouldn't renew my lease.

Had she said such a thing about Casey, I wouldn't have given moving a second thought. But was I willing to give up a great apartment for two creatures I never really wanted in the first place? Though I didn't think I loved my cats, I did feel for them and increasingly enjoyed their company. They were growing on me.

I agonized over the decision. I asked my friends for advice, but their counsel depended entirely on whether they were dog or cat people. Very few of my friends are both, even though chef Dereke Bruce may have been on to something when he said, "In order to keep a true perspective of one's importance, everyone should have a dog that will worship him and a cat that will ignore him."

(Studies have found significant personality differences between dog and cat lovers. A team of psychologists led by Sam Gosling at UT-Austin conducted a web-based study of nearly five thousand people they divided into groups based on whether they identified as dog people, cat people, both, or neither. Dog lovers scored higher than all three other groups on both agreeableness and extraversion; they essentially tied with nonanimal people for the lead on conscientiousness; and they scored the lowest — which, in this case, is actually the best — on neuroticism. The only category where dog people didn't fare as well was on "openness," which Gosling defined as appreciating art, emotion, unusual ideas, and a variety of experiences.)

Some of my dog friends encouraged me to get rid of Eli and Chi. My cat friends scolded me for even entertaining that idea. "Benoit, I'll be *very disappointed* in you if you do this," one insisted over sushi. "You made a commitment to these animals."

He was right. I had made a commitment. And, when it comes to dogs, I'm horrified by the stories I've heard from shelters and animal rescue organizations about dog owners who relinquish their pets because an apartment they covet doesn't allow animals — or, worse yet, because their dog's shedding hair doesn't match the color of their new carpet. How could people be so heartless? How could they live with pets, and then just get rid of them?

In the end, though, that's what I did. I got rid of the cats. To my credit, I found a family that wanted to adopt them together. But almost as soon as I agreed to the exchange, I started having second thoughts. When the family — mom, dad, and two kids — walked in my front door to claim their pets, I wanted to tell them to leave. Though I couldn't put my finger on it, they seemed *off* somehow. Was I making a terrible mistake?

My gut urged me to slam the door in their

faces, but I still handed Eli and Chi over. I immediately felt like the worst person in the world, and as the family drove away with my cats, I surprised myself by walking to my bedroom, collapsing on the bed, and bursting into tears.

The Cherry Hill RV Park in College Park, Maryland, couldn't have been more different from Liberty Harbor. A popular summer destination for families, Cherry Hill boasts swimming pools, ponds, an outdoor theater, a game room, and a hiking trail. It felt less like an RV park and more like summer camp.

It also featured a grassy, fenced-in dog play area where I took Casey to pee and play. I laughed when he ignored several skinny tree stumps, urinating instead on a fire hydrant that had been placed in the run as a joke. Fake hydrants are a fairly common sight in dog parks and pens, but they come with a potential downside: volunteer firefighters in Oak Bluffs, Massachusetts, once lost valuable time when they tried to hook up a hose to a nonworking hydrant in a dog pen near a burning home.

Other than their primary utility (putting out fires), hydrants in high-traffic dog areas function as a kind of Facebook for dogs, according to Alexandra Horowitz of the Horo-witz Dog Cognition Lab at Barnard College. She says urine on a hydrant can advertise everything from a dog's confidence to where he's from.

"In this way," she wrote in her 2009 book, *Inside of a Dog,* "the invisible pile of scents on the hydrant becomes a community center bulletin board, with old, deteriorating announcements and requests peeking out from underneath more recent posts."

Our first full day in Washington was clear and unseasonably warm, so I decided to walk Casey on the National Mall. Signs warned me to keep my dog on a leash, but I pretended not to notice them and played fetch with Casey on the grass at the foot of the Washington Monument. Every once in a while, Casey would stop running and roll on his back in the grass as American flags snapped in the wind above us.

I bought a hot dog from a corner stand and led Casey toward the White House. I hoped to get lucky and catch a glimpse of the First Dog, Bo, but the presidential pup was out of sight. I imagined him dozing on an overpriced doggie bed in the Oval Office.

During Bo's official introduction to the media in April of 2009, President Obama had made clear that the dog would be welcomed everywhere in the White House. "I finally got a friend," Obama said at the time, invoking Harry Truman's famous quote, *If you want a friend in Washington, get a dog.*

Just about every U.S. president has heeded that advice. (Presidential cats, on the other hand, tend to stay out of sight. "I'm not sure you could get elected these days as a person who owned a cat but no dog," wrote *New York Times* columnist David Brooks.) The First Dog dates back to the inception of the presidency itself, even before the construction of the White House. George Washington, who is credited with the creation of the American Foxhound, kept dozens of hounds at his Mount Vernon estate. At least thirty, including one named Drunkard, are mentioned by name in his diaries.

Washington understood how much a dog could mean to his fellow man — even his nemesis. When a Fox Terrier wandered into Washington's Pennsylvania camp during the Revolutionary War, his men considered keeping the dog as a kind of mascot. But when one of them noticed the name engraved on its collar (General Howe, the British commander), Washington insisted

153

they return the dog to its rightful owner "under a white flag of truce."

When John Adams moved into the White House a few years later, he brought two mischievous mutts along with him. He named one Juno, after the fierce Roman goddess. He named the other Satan, a name that probably wouldn't fly in today's political climate.

Adam's dogs were the very first in a long, nearly uninterrupted succession of canine residents of the White House. To not have a dog — or many dogs, as was the case with several presidents, including Calvin Coolidge and John Kennedy — was considered politically risky.

Thomas Jefferson, for example, can be said to have flip-flopped on the dog issue. In response to a friend who wrote to him complaining of stray dogs, Jefferson had once declared, "I participate in all your hostility to dogs and would readily join in any plan of exterminating the whole race. I consider them the most afflicting of all follies for which men tax themselves." Yet he kept many on his estate and is credited with introducing the Briard breed to North America.

But perhaps no First Dog has received such intense media scrutiny as Bo, the

Portuguese Water Dog owned by the Obamas. During his victory speech in November 2008, Obama promised his daughters that a puppy would join the family when they moved into the White House. But Malia's allergies necessitated a dander-free dog.

Even before Obama took office, the question of breed became a public debate. While an AKC online poll revealed that most Americans wanted the family to buy a Poodle, others scolded Obama for even considering a purebred. When he selected one anyway, the criticism was predictably fierce. "This will fuel the breeding industry, which will fuel the puppy mill industry, which will increase homeless dogs at shelters," Dr. Jana Kohl, author of *A Rare Breed of Love,* said at the time.

But Obama looked great in comparison to his 2012 Republican challenger, Mitt Romney, who was widely portrayed as an animal abuser for once strapping his Irish Setter to the roof of the family car in a makeshift kennel for the twelve-hour drive from Boston to Ontario.

When I spoke to Democratic strategist Doug Hattaway two months before the election, he said Romney's dog gaffe "plays perfectly into the narrative of him as an aloof, uncaring, out-of-touch rich guy. A lot of people already think he doesn't care about poor people, but it's probably even worse for him to be seen as someone who doesn't care about *dogs.* You don't want to mess with dog-loving voters."

When it comes to the treatment of animals in this country, few people hold more power than the woman Casey and I went to meet on our second day in D.C.

Ingrid Newkirk looks like a cross between a German nanny and a retired professional tennis player, but she's devoted her life to a different kind of sport — the rough-and-tumble game of animal rights. She founded and rules over arguably the most successful and controversial animal rights organization in the world, People for the Ethical Treatment of Animals (PETA).

Best known for their attention-seeking tactics, PETA activists have hurled red paint at fashion models and tofu-cream pies at designers, donned KKK outfits in a protest of Westminster, and dropped a dead raccoon on the lunch platter of *Vogue* editor Anna Wintour. Before visiting Ingrid, I'd asked my Facebook followers what they thought about PETA. The response was immediate — and damning.

"They hurt the cause of animal rights and hurt it badly," one woman said. Another lamented that "they started out with a good purpose but let their egos send them into radical fanaticism." One reader didn't mince words, calling the organization's leadership "terrorists." (Comedian Dennis Miller once joked that "there are even animals out there who are embarrassed that [PETA activists] front them.")

Many animal rights groups have spoken out against PETA, condemning it for hijacking the movement and trivializing the suffering of animals. And, yet, even with its battered reputation, PETA remains a fundraising juggernaut that has scored numerous victories for animals. Some of the biggest names in clothing and fast food have bowed to pressure from Ingrid and her relentless activists.

Ingrid runs her organization like a dictatorship — "This is not a democratic organization," she told *The New Yorker* in 2003 — and I expected to be faced with a firebrand when we sat down in her sparse office on the second floor of a charming white-brick building near Dupont Circle. (It was a Saturday, and we were the only ones at the office.) But Ingrid offered me a cup of tea and proceeded, even when I disagreed with her, to mostly come off as personable and friendly.

Though I'd come to talk to her about dogs, I began our ninety-minute chat by asking why PETA seemed willing to do anything for attention. In the *New Yorker* profile, she openly referred to PETA as "press sluts." Why, I wanted to know, did PETA so often seem like a kid with a temper tantrum?

"If you're a social cause in this country," she told me, "you're basically dead unless you do something titillating, something controversial, something that's seen as extreme. We will use whatever tactics we can think of to get people talking about an issue."

"But what if your tactics turn people off from the issue you're fighting for?" I countered. "Many people can't see past your tactics."

"To me that's hilarious," she said, though she didn't seem amused. "Animals are being skinned alive for fur, they're being bludgeoned in leg-hold traps, and the pigs that make a ham sandwich have black lung and are lying in filth. *Those* tactics are okay, but us saying 'Would you eat your dog?' " — a reference to a PETA billboard campaign aimed at getting kids to stop eating

158

turkey — "is somehow upsetting and over-the line? PETA annoys people because it gets under their skin. It makes people uncomfortable."

I turned the conversation to PETA's unusual perspective on dog ownership. In Ingrid's conception of a perfect world, humans would not own dogs or have the ability to purchase them. On the organization's website, an entry on the issue reads: "We believe that it would have been in the animals' best interests if the institution of 'pet keeping' — breeding animals to be kept and regarded as 'pets' — never existed. . . . This selfish desire to possess animals and receive love from them causes immeasurable suffering, which results from manipulating their breeding, selling or giving them away casually, and depriving them of the opportunity to engage in their natural behavior. They are restricted to human homes, where they must obey commands and can only eat, drink, and even urinate when humans allow them to."

Ingrid conceded to me that a world without pet keeping is unlikely, and that her own childhood would have been markedly different without her best friend, an Irish Red Setter named Shawnie.

"Growing up, my dog was like my

brother," she recalled. "Shawnie and I slept in the same bed. We threw up in the car at the same time, because we both got miserably carsick. If I had ice cream, he would eat the top of it. We were the kids of the family."

Though Ingrid said she travels too often to live with a pet today, she encourages employees to bring their dogs, cats, and other "companion animals" to work with them. And, in a nod to the powerful bonds inherent in a world of pet keeping, PETA offers bereavement leave — time off for the death of an employee's pet.

But though she expects her employees to treat dogs well, she insists that most humans can't be similarly trusted. "The institution of pet keeping has resulted in six to eight million dogs and cats being abandoned in animal shelters and pounds in the U.S. alone every year," she told me. "That's a whole lot of confused, sad, old, injured, unsocialized animals who find themselves on a cement floor in a noisy kennel or cage, and at least half of them will have to be destroyed because people are out buying more animals from breeders and pet shops. And then there are the thousands of dogs who supposedly have homes, but who are neglected — left in their pens, or chained up in backyards. So, while we all see lovely, happy dogs being walked, the reality is that the practice of pet keeping is responsible for a truly awful situation for the very individuals we love."

I asked Ingrid what she might replace the institution of pet keeping with. "A system where animals live in peace without being acquired to amuse us and be dependent upon us," she replied. "Humans should only take in animals who are in need — orphaned, injured, or starving. But only animals who genuinely need a human helping hand."

I agree with Ingrid that too many American dogs are failed by their caretakers, just as too many American kids are failed by their parents. But she was reluctant to acknowledge that some dogs might be better off — might even *prefer* — sharing their lives with humans.

PETA argues that dogs should be allowed to "engage in their natural behavior." But for the past several millennia, the "natural behavior" of dogs has been to live alongside us. For the domesticated dog, human society *is* its natural habitat. Dogs rely on humans for their survival as much as we rely on them for companionship. This "romantic and timeless . . . interspecies contract," as writer Lars Eighner once put it, is somehow lost on PETA.

Ingrid insists that even dog owners with what she calls "the best of intentions" don't always understand what a dog actually needs. She ticked off a handful of examples: pet owners who don't think to move an arthritic dog from a draft; who believe that a daily walk or two around the block is enough exercise for a young dog; who don't stop to let a dog sniff whatever it wants; and who don't let a dog bark freely.

"Barking is totally natural for many dogs, but too many people yell at them for it," she said. "You see them on the street screaming, 'Be quiet! Be quiet! Do this, do that.' As if the dog is in the Marines."

I thought about all the times I'd impatiently shooed Casey away from something he wanted to sniff, or snapped at him to "stop barking" when I wasn't in the mood. Was it cruel of me to not let Casey speak whenever he wanted? Is unfettered barking really what a dog needs?

As I listened to Ingrid, it seemed to me that her conception of a good dog life is one where the animal never hears the word "no." But is a life without rules, structure, or delayed gratification ideal for any animal, canine or human? As much as most of us want to protect our pets from suffering, I couldn't help thinking of dog trainer William Koehler, who had "the now old-fashioned idea that suffering might have an important place in the good life — even for a dog," John Homans writes in *What's a Dog For?*

Ingrid went on, calling out people who leave their dog home alone for too long. "I tell people, 'If you expect your dog to hold it for eight hours, then *you* hold it for eight hours at work,' " she said. "See how that feels. Because dogs don't have any supernatural powers."

She's especially frustrated by those who treat their pets like "an accouterment" to their lifestyle. "A dog is not a handbag," she said, handing me a pink and green coffeetable book she wrote called *Let's Have a Dog Party! 20 Tail-Wagging Celebrations to Share with Your Best Friend.*

Ingrid explained that the book looks frivolous by design — she's hoping to reach the Paris Hilton crowd, she told me. Interspersed between chapters on hosting fabulous dog parties (including "bark" mitzvahs) and creating doggie goodie bags, Ingrid imparts serious messages about what she considers responsible dog ownership, from buckling up a dog in a car to letting him run freely in a park.

Toward the end of our talk, I brought up what I would soon realize was a touchy subject — the euthanasia rate of PETA's shelter at its headquarters in Norfolk, Virginia. In 2011, PETA euthanized 95 percent of the 2,050 dogs that entered its shelter. In the last fifteen years, PETA has put down some 27,000 animals.

For an organization that claims to care so much about them, I couldn't understand those numbers. When I pressed Ingrid on them, her face tensed for the first time. "We are a shelter of *last resort,*" she told me, sounding exasperated at having to explain herself. "We take the broken animals that we know from experience no one wants, the animals that people are giving up because they're elderly and they can't afford the vet fee to put them down. We take in the dregs. These are mostly unadoptable dogs."

I had a hard time accepting that answer. I told Ingrid about Randy Grim, a man who has devoted his life to saving unwanted and supposedly unadoptable dogs in the St. Louis area. In 2009, I'd spent a week with Randy rescuing dogs — dregs, all of them — from two dangerous, dilapidated neighborhoods. (I was looking forward to visiting him again on this trip.) With the help of his small staff and a volunteer army of dog lovers, Randy has proven that very few dogs are truly unadoptable.

"I know that there's nothing more emotionally satisfying than putting your arms around a dog and saying, 'I saved you,' " Ingrid said. "Anyone who does anything to help is appreciated, but touchy-feely doesn't alter the scheme of things. Wouldn't that money and energy be better spent on a spay and neuter program so that we can stop the birth of many dogs who are born under the steps of a trailer and who will have no shot at a good life?"

Ingrid told me that PETA spays about ten thousand dogs each year at no or low cost. It also focuses on trying to pass ordinances banning chaining and on counseling dog owners before they bring their animal to a shelter.

"If anyone who criticizes our priorities wants to come to our shelter and adopt one of these dogs, then come on down," she said. "We're doing what we think will do the most good for the most dogs."

It's no surprise, really, that PETA's shelter kill rate is so high. If you believe, as Ingrid does, that too many dogs suffer needlessly at the hands of human caretakers, then what better way to alleviate that suffering than by permanently *ending their suffering*? A dead dog can't be chained outside, or cooped up all day inside, or treated like a handbag.

But a dead dog is also cheated out of a chance to be paired successfully with humans, nor can it bring meaning and joy to people. PETA might dismiss this kind of thinking as "our selfish desire to possess animals and receive love from them," but that's just one of the ways that the group caricatures — and sometimes misunderstands — the human-canine bond.

I looked down at Casey, who napped on the floor in Ingrid's office. Though I'd spent many of our years together worrying that he might rather be paired with a different human, I had no doubt that he wanted — needed — my species.

I wondered what PETA would do if Casey meandered alone with no tags into its Virginia shelter. Would its workers look at my nine-year-old dog and assume that he's been failed by the human race? Would they huddle up and agree that it would be a waste of time and resources to try to find Casey a human who wouldn't fail him? Would they convince themselves that the most humane course of action would be to kill my dog?

■ ■ ■ ■

On our final morning in the D.C. area, Casey and I went for a long walk through the Cherry Hill Road Recreation Center, a park that connects Cherry Hill with the University of Maryland.

Casey trotted along ahead of me, a thin tree branch in his mouth. When a black cat darted across the path ahead of us, he dropped the stick and halfheartedly chased the animal into some bushes.

"You're not going to catch anything with that kind of effort," I said, forgetting, as I often do, that Casey is a dog. (Patricia McConnell writes in *The Other End of the Leash* that the human need to talk to dogs is so strong that dog trainers "talk to deaf dogs even when we know they can't hear us.")

On our way back to the RV park, we came upon a tiny woman being walked by a large German Shepherd. The woman did not look happy. She was panting heavily and leaning backward with all her body weight, trying to reel the dog in as if it were a shark. I feared the leash might snap.

"Big boy you have there," I said as the German Shepherd came to a full stop to

167

check out Casey. The woman breathed a sigh of relief.

"If I let him go," she told me sheepishly, "he'll run away. He doesn't listen to anything I say."

The dog was her son's, she explained, but the teenager had recently escaped to college. "I guess it's just *assumed* that mom gets stuck taking care of the dog," she said. "It's like everything else we get stuck doing. No one asks for our opinion!"

"A grave injustice," I agreed.

She pulled a water bottle from her coat pocket. "What's your dog's name?"

"Casey," I replied. "I know — not very original."

"Oh, this guy's name is even more boring," she said. "My son named him Max."

Though Max has been the most popular male dog name for several years, I couldn't remember meeting a dog named Max before. They must run in different circles, I thought. But I did know a Max-related dog story. I told the woman — as is typical of these encounters, I learned her dog's name but not hers — about Gestalt psychologist Kurt Koffka, who bestowed the name Max to all seven of his Dachshunds. After Koffka's first dog named Max died, he got another dog and initially called him a different name.

"Yet he looked like Max, and he acted like Max, and sometimes I found myself calling him Max," Koffka recalled. "So I said to myself, 'If he wants to be Max, then he is Max.' I suppose that I just wanted him to be Max and to live forever. That is the nice thing about a purebred dog. If you find one you like, you can have him again and again, since they are all so much the same. I suppose I call them all Max because they *are* all Max, and that is who I wish to live with."

I confessed to the woman that I harbored some shame about naming my dog Casey. The word held no special significance for me, nor was it in the least bit clever. It had popped into my head a few hours after meeting Casey as a puppy. It had seemed to suit him, so I went with it.

Dogs, of course, probably don't care what we call them. In fact, dog trainer Brian Kilcommons likes to joke that most American dogs probably think their name is "No!" Or, in the case of a dog who lives in a house with thirteen cats, "Kitty." Pet photographer Walter Chandoha named his dog Kitty to simplify mealtime. When Walter calls out, "Here, Kitty Kitty," thirteen cats — and one dog — come running from all directions.

In the book *The Complete Book of Pet*

Names, Roger Caras writes in the foreword that "a name for a pet is, or at least should be, equally meaningful" as the name given to a human child. But Caras concedes that there are important differences in naming kids and dogs.

"Whimsy and humor can generally play a somewhat larger role than they usually do with our children's names," he writes, surely aware that it would be cruel to name a human child Mayhem, Trouble, Sassy, or Bear. (Dogs don't have to survive middle school, after all.) But Caras urges dog owners to reach deep into the vault. "The naming of a pet is still a creative act," he insists.

The best pet name I'd come across on my journey to that point belonged to a Newfoundland/Chow mix I met at Tompkins Square Park. The dog's owner, a mystery buff, named him Alibi. Sometimes, a deceptively straightforward name can be equally brilliant. A friend of mine calls his Jack Russell mix Little Dog. When the dog scampers out of sight for too long, my friend pipes up and yells, "Here, Little Dog!"

There's something to be said for naming your dog a word that's fun to shout. I heard about a high school English teacher who named her dog Spot so she could quote Macbeth ("Out, damned Spot!") every time she let the dog into the yard. Another woman — this one *A Streetcar Named Desire* fan — called her dog Stella for the sole purpose of leaning out her back porch and screaming, "Stella! Stella!" Then there's the cousin of a friend who named his dog I Am. He was apparently in a long-standing feud with his neighbors, and it pleased him to stand on his porch and call his dog by yelling, "Here I Am!"

In this day and age, when pet names increasingly overlap with human ones, calling out to your dog can lead to some confusion. My French cousin told me that his neighbor's dog has the same name — Jacques — as my cousin's father.

"The dog hangs out in our garden a lot," my cousin explained, "so it's hilarious to hear the neighbor scream out with authority, 'JACQUES! COME HERE!' My dad's always confused for a second, because he doesn't know if it's my mom calling him, or the neighbor calling the dog."

When I asked my friends and Facebook followers how they'd chosen names for their dogs, I was impressed by their due diligence. Many had spent days or weeks selecting the perfect name. Some were careful to match it to the dog's personality or physical attributes. A man named his shy Labradoodle

Shy. A woman named her dog Jack Sparrow (from *Pirates of the Caribbean*) because it was "one-eyed and as cute as Johnny Depp."

As I tabulated dog names, several trends emerged: a surprising number of dogs are named after Grateful Dead band members and songs — Aiko, Ripple, Jerry Garcia, Dupree. Countless dogs have sports-themed names — a Green Bay Packers fan, for instance, named her dogs Packer, Green Bay, and Lambeau. If there's one book that seems to inspire dog lovers, it's *To Kill a Mockingbird.* Scouts and Atticuses are everywhere.

Among art lovers and feminists, the name Frida (after Mexican painter Frida Kahlo) is a popular choice. And among those who rescue their dogs off the street, it's common to name the animal for where it was found. A dog discovered in a poison ivy patch becomes Ivy; one rescued from a freeway becomes Freeway.

I also discovered some breed-specific tendencies. Those who live with Pit Bulls often give their dog the most nonthreatening name possible — Bella, Sophie, Angel, that kind of thing. Many Basset Hound owners believe their dogs need "old people" or "hillbilly" names.

"I wanted to call my Basset Hounds Andy

and Barney," one woman told me, "but they were girls so we ended up with Miss Jane and Ellie Mae."

Kids, I was reminded again and again, tend to complicate the dog-naming process. "Never, *ever* let a child name your dogs," a friend told me. "You'll end up with a bunch of obscure *Aladdin* characters running around your house!"

One mother recounted rejecting most of the dog names suggested by her seven-year-old twins, including countless *Star Wars* characters and the *Harry Potter* character Nymphadora. They finally settled on Doo, named after the *Harry Potter* house-elf Dobby.

For couples that try on a dog before having children, I learned that it's wise to pick your naming battles. One woman relented to her husband's wish and called her yellow Lab Rocco, "to avoid the possibility of ever having to give our son that name."

Though naming a dog can seem like fun and games, it can also be serious business. At animal shelters across the country, employees work hard to come up with likable names for their dogs — especially ones stuck in the shelter for weeks or months. I heard about one shelter that named an entire litter after characters from the TV show *Friends.*

"Where I used to work, all these dogs would come in with unattractive names that really wouldn't do them any favors," said Rodney Taylor, the associate director of the Animal Management Division in Prince George's County, Maryland. During my visit to his shelter, Taylor told me that he increasingly instructs his staff to give the dogs celebrity names; recent choices had included Foxy Brown, Faith Hill, Al Capone, and Beyoncé.

At the shelter, I learned that an enticing name is especially important for black dogs. "Many people subconsciously overlook them," explained Amanda Leonard, an expert in what many shelter workers call Black Dog Syndrome, a phenomenon where dark-coated dogs are ignored in favor of their lighter-coated counterparts.

Amanda, who joined me during my trip to the shelter, had first become interested in Black Dog Syndrome when she worked at the Washington Humane Society and noticed that black dogs were the last to be adopted and the first to be euthanized.

In a later paper on the subject published in UC Berkeley's *Kroeber Anthropological Society,* Amanda wrote that black dogs can appear more menacing than they are. Black absorbs light, which moderates a black dog's facial feature definition and "makes it harder to read their facial expressions." Since there's little differentiation between black eyes and a black coat, it's difficult for adopters to gauge a black dog's mood, or to make an emotional connection with the animal.

For the same reason, black dogs are a challenge to photograph well — their features become obscured by their dark coat. "Often, the crucial first glimpse a potential adopter has of a dog is through the shelter's website," Amanda explained. "If the picture of a dog is not good, many adopters will move on."

That meant bad news for several black dogs I met at the Prince George's shelter, including Oliver, a three-year-old black Lab mix who had been found by police running loose in a park. He's been at the shelter for more than a month.

"He's a great dog — he's friendly, and he's one of the staff favorites," explained a young shelter worker who showed us around.

"So why hasn't he been adopted?" I asked

"Black Dog Syndrome," she guessed.

We also spent some time in front of the

kennel of a two-year-old Rottweiler mix, who took some coaxing before letting us pet him through the grate. Amanda said that the dog — who was surrendered to the shelter on Valentine's Day — would be difficult to adopt because he's shy, black, and positioned next to an outgoing white dog.

Most shelters are aware of Black Dog Syndrome, and they do their best to make their darker dogs stand out. Some hire professional animal photographers to capture the dogs' personalities. Others dress the dogs up in colorful bandannas or ribbons. Shelters will also avoid kenneling multiple black dogs next to each other, so they don't appear commonplace, and will train dogs to come to the front of their kennels when prospective adopters approach. Some even organize black dog promotions, reducing adoption fees for black animals.

Amanda was careful to stress that most people don't consciously discriminate against black dogs. "It's not like people come into a shelter saying, 'I am *not* leaving here with a black dog!' " she told me.

Amanda speculated that black dog bias might have a cultural origin. In British folklore, black dogs are often portrayed as a sinister omen of death, most famously as the hellhound in Sir Arthur Conan Doyle's

The Hound of the Baskervilles and the Grim in J. K. Rowling's *Harry Potter and the Prisoner of Azkaban.* The origins of the ominous black dog (black cats are also less likely to get adopted) may reach all the way back to ancient Greece — Plutarch's biography of the prominent Athenian statesman Cimon mentions that he dreamed of an angry dog barking at him, which foretold his impending death. (And remember the Black Dog of the Hanging Hills?)

Winston Churchill famously referred to his bouts with depression as his "black dog," a constant lurking companion that he could neither master nor rid himself of. Even the color black itself carries a negative symbolic association, which some argue could subconsciously prompt potential adopters to pass on black dogs.

After our morning walk through the Cherry Hill Road Recreation Center, I led Casey back to the Chalet and prepared the cabin for departure. We had three stops to make that day in Virginia. Or, I should say, we had three stops to make that day in *northern* Virginia. The state's northern and southern populations make for strange bedfellows.

First on the agenda was a meeting in Fairfax with Kimberly Zakrzewski, a stay-at-home mom who was dragged to court for allegedly breaking her county's pooper-scooper law. From there we'd continue west and treat ourselves to an afternoon at Barrel Oak, one of an increasing number of dog-friendly wineries in America. We'd finish the day at the Flint Hill Public House and Country Inn, a pet-friendly hotel in Flint Hill, Virginia. The owners had heard about my journey and offered a free night's stay.

After two weeks in the Northeast with very little actual driving, Casey and I were now embarking on what felt like a real road trip. This was not good news for him. The next few weeks would test my dog's tolerance for life on the road — we'd be rolling through Virginia, North Carolina, South Carolina, Georgia, and Florida. To my credit, I tried to remedy the situation at The Big Bad Woof, a cleverly named pet store in Hyattsville, Maryland, that boasts "essentials for the socially conscious pet." When I explained Casey's predicament to the woman behind the counter, she suggested a frozen lamb shank and some homeopathic "calm down" medicine for dogs.

She pointed to a shelf with bottles of drops and chewable tablets with names like Tranquility Blend, Travel Calm, Pet Calm, and Emotional Balance. I picked one at random (Travel Calm, by Vet's Best) and noted that one of its ingredients, tryptophan, has long been associated with post–Thanksgiving turkey comas. Another ingredient, valerian root, is "purported to possess sedative and anxiolytic effects," according to my quick iPhone Google search in the store. I'd never heard the word "anxiolytic" before, but I assumed it was a close cousin to anxiety. And Casey had plenty of that in the RV.

Still, I was skeptical that Travel Calm would solve the issue. "I don't suppose you sell dog tranquilizer guns?" I joked.

Back at the motorhome, I fed Casey two chewable Travel Calm tablets, as instructed. To pass the time while I waited for the "sedative and anxiolytic effects" to kick in, I did pushups and biceps curls on the grass outside the RV. I'd been concerned about nearly four months away from the gym, so I'd purchased adjustable dumbbells and asked my trainer to develop an "RV Workout plan."

Though my trainer spoke romantically about my outdoor workouts at the foot of the Rockies, his plan was weighted heavily toward lunges and other laborious leg exercises. Those weren't going to happen;

179

the only reason I do them during our normal training sessions is because he's there to berate me. Minus kidnapping my trainer for this journey, there was little chance I'd follow his recommendations. In fact, I'd already misplaced the paper on which he'd outlined them.

After some painless biceps curls, I noticed that I'd missed a call from Dr. Gold, the therapist I'd visited in New York City. On my voicemail he asked me how the trip was going and added that I should feel free to call him.

"I know our meeting brought up some feelings for you," he said. "I'm here if you want to process anything."

I smiled at the sound of his voice. Though I'd thought a lot about our time together in the day or two after it, the stress and excitement of my first two weeks on the road mostly crowded out any introspection. But hearing him that morning made me stop and take stock. How *was* the journey going so far? How *did* I feel about my relationship to Casey? Had anything *changed*?

The answers were muddled. On one hand, I'd felt something begin to shift almost immediately after leaving Dr. Gold's office: I started personalizing Casey's behavior less. He's a dog, after all, and it's not his job to

180

care for my emotional needs. As I began to accept that, I was able to notice — and, occasionally, laugh at myself — when I caught myself interpreting his aloofness as proof of my unlovability.

I also started paying more attention to what *he* might need. There's a time most days (usually in the evening) when Casey wants to be petted, wants a minute of closeness and quiet time. But he doesn't just shove his head in your lap like some dogs. Instead, he'll stand or sit a few feet away from me, his face relaxed and content, and wait for me to read his mind.

It's one of the only subtle things Casey does, and if my mind is elsewhere I'm guaranteed to miss it. Other times — if I'm too busy resenting Casey for not being "the dog of my dreams" — I'll pretend to miss it. I'd confessed this to Dr. Gold during a shame-spiral in his office.

"What kind of terrible person acts passive-aggressively toward his dog?" I'd said. "What kind of person withholds love from a pet?"

During our session, Dr. Gold had helped me make a connection between my behavior toward Casey and my mother's behavior toward me. My mother was a master of withholding affection, of not picking up on my cries for attention. Even my dad, who eagerly expressed his love, wasn't aware of the extraordinary sadness and loneliness I felt as a kid.

With the RV ready for the road, I looked at Casey. If anything, he seemed more awake than before chewing the Travel Calm tablets. Even the lamb shank didn't do the trick. My food-obsessed dog forgot all about it ten minutes into our drive, when a gust of wind pushed the RV onto the rumble strip. He dropped the shank and walked — head low, shoulders slumped — from under the dinette to the small space between the driver and passenger seats.

"At least he wants to be close to you," a friend said when I called him from the road.

"That's not the reason he moved," I assured him. "This is just the smallest, most secure spot in the RV."

I laughed as I heard myself. "I think I might need more therapy."

Kim Zakrzewski was recovering from the flu when I arrived in Fairfax, and she joked — not all that inaccurately — that she looked near death. She's also a chain-smoker, which only added to her aura of gloom.

Soon after we arrived at Kim's modest

182

condo community, Casey squatted on some grass to poop. Before I could retrieve a doggie bag from my coat pocket, Kim, who wore a black jacket with a faux fur hood, handed me one from hers.

"Wow, you're quick," I said.

"Honey, I've got poop bags coming out of my ears!" she cackled.

Kim retrieved Baxter, a white Westie/Bichon Frise mix, from her apartment. (The dog isn't hers, but she walks it most days for a friend who lives in the same complex.) It was Baxter's poop — and Kim's alleged failure to dispose of it — that made her a household name in Fairfax.

"All the papers and television stations covered my trial, so now I go to the bank and the teller says, 'Oh my God, you're the poop lady!' " Kim told me as we walked our dogs around her complex. It was a cold, poopy day, and Casey's tail wagged crazily as he came upon remnants of what appeared to be a long-dead squirrel.

(Belgian-born poet Henri Michaux correctly remarked that dogs never stop to smell a rose. "Who understands the menu of stink better?" he wondered in his 1950 book, *Passages*. "These innocents come back to our sides without skipping a beat, full of affection and radiating a clear conscience.")

Kim pointed to a second-floor window next door to her unit. "That's where *they live,*" she whispered. By *they,* Kim meant sisters Virginia and Christine Cornell, who in 2011 snapped photographs from that very window in an attempt to prove that Kim had failed to pick up Baxter's droppings, as mandated by Fairfax County's pooper-scooper law.

The case ended up in court, where a jury of Kim's peers listened as the sisters accused her of numerous indignities. According to an interview they gave to *The Washington Post,* Kim would let "the dog poop on purpose because she knew it annoyed us" and had "no respect or regard for anyone else and views herself as above the law."

The sisters produced photographs of Kim walking Baxter, as well as images of the allegedly offending poops, which they'd taken while surreptitiously tailing Kim and Baxter from the cover of trees and bushes. Christine testified that during their surveillance, "not once during these three days did [Kim] bend down or did she produce a doggie bag," adding that "every time she has the white dog, poop appears around our building."

The defense countered that while the prosecution had photographs of Kim walking the dog, as well as pictures of unidentified dog poop, they had no pictures of Baxter *actually pooping*. (The words "poop" and "pooping" were bandied about in court. "Will it offend you if I use the word 'poop'?" the commonwealth's attorney asked the jury early in the day.)

During cross-examination, Kim's lawyer pressed Christine about whether she had "any pictures of the dog going to the restroom."

"Do dogs really use *the bathroom*?" the judge interjected, causing snickering in the courtroom.

The attorney soldiered on. "How do you know the pile of poop you took pictures of was the one [Baxter produced]?"

"You can tell fresh dog poop," Christine insisted. "It's that simple. Old poop dries. There was a *moist* appearance to it."

There was considerable testimony about Baxter's diet, with the defense contending that the droppings in the photographs were much too large to have been produced by a nineteen-pound dog.

"Looks like a horse's poop," Baxter's owner, Michelle Berman, testified. "There's no way that came out of my dog."

Michelle offered to present a sample of Baxter's droppings for comparison (she had some in her car), but the judge quipped that Michelle probably wouldn't get it past security.

As the day went on, the jury learned that this poop-related dispute wasn't really about poop at all. Kim and the sisters had been feuding for years. Police had been called to the complex a handful of times over accusations of slashed tires and damaged doormats, and in 2008 the sisters had filed an unsuccessful complaint charging Kim with reckless driving. They claimed she had tried to run them over.

In the end, the jury in the poop case took only twenty minutes to side with Kim. "I joke about what's happened," Kim told me, "but I still get panic attacks about living here. People tell me that I should just move, but this is my home, and why should I let them run me out?"

As we spoke, two women scurried across the complex parking lot to their car. "Oh my God, that's them," Kim whispered again. Casey pulled at the leash; he wanted to say hello to the sisters.

"Trust me, they don't like your kind," Kim told Casey with a laugh. She turned to face me as the sisters ducked into their car.

"They'll probably drive really slow to check you out."

Kim was right — the sisters craned their necks to get a good look at us as they rolled through the lot. I would have liked to get their side of the story, but they hadn't responded to my requests for comment.

Kim's apartment complex isn't the only one in Virginia with poop issues. In its coverage of the case, *The Washington Post* pointed out that two other area properties had taken a hard line against scoop-defiant dog owners. PooPrints, a service of BioPet Vet Lab in Tennessee, allows building property managers to take a DNA swab of each dog living in a complex. When droppings are found on property grounds, they're sent to the lab and matched to the offending dog.

According to BioPet (which, to be sure, has a dog in this fight), 38 percent of canine waste in America is waiting to be stepped on, or absorbed into the atmosphere. One study of air quality in Midwestern cities like Cleveland and Detroit found that "fecal matter, most likely dog feces, often represents an unexpected source of bacteria in the atmosphere at more urbanized locations during the winter."

But irresponsible dog owners aren't confined to the Upper Midwest. Even John Zeaman, the New Jersey writer who authored *Dog Walks Man,* admits to being part of the problem. "I don't always pick up," he bravely confesses in print. "I know this raises character questions. My wife will be horrified that I am making such a public admission."

After all that poop talk, I needed a drink. Fortunately, we weren't far from Delaplane, Virginia, which claims to have more wineries than any other zip code on the East Coast.

I was skeptical of Virginia wine. Truth be told, I'd never heard of Virginia wine. Raised in San Francisco by a French mother, I grew up on imports from Bordeaux and the Rhône and Loire Valleys, and local reds from nearby Sonoma and Napa.

I pulled the RV into Barrel Oak's 270-acre property, on which sits the boyhood home of former chief justice John Marshall. On a gravel road, we passed a sign that set the tone for our afternoon: Slow Please — Dogs, Kids & Winemakers at Play.

Casey and I bounded out of the motorhome and were met by a friendly eleven-year-old Hungarian Vizsla named Birch. The slender brown and white dog "just showed up one day," according to Sharon Roeder, who opened Barrel Oak in 2008 with her husband, Brian, a big bearded man who looks just like actor Bruce McGill. (McGill is best known for playing the role of Jack Dalton in *MacGyver* — and, when he was younger, D-Day in *Animal House*.)

"We think Birch was a farm dog from around here who was used to fending for himself," Sharon shouted in the winery's tank room over the roar of a power washer used to clean empty wine barrels. "He would come over and hang out on the patio and play with the dogs and mooch people's picnics, though I haven't found his snout in a wineglass yet."

Brian and Sharon used to shoo Birch out at night, but at one point he just stopped leaving. "He totally adopted us," Sharon said.

In Barrel Oak's spacious tasting room, Brian poured me a glass of the winery's 2010 Chardonnay Reserve, which only months earlier had received the highest prize at the *San Francisco Chronicle* International Wine Competition. From my spot at the tasting bar, I could see a black Lab lounging by the fire and a Beagle pitter-pattering around his owner's booted feet near the patio. Across the room, Casey and

Birch jockeyed for position on a leather couch near a wooden beam with signs that read, *Dogs have many friends because they wag their tails instead of their tongues,* and *If Cats could talk, they wouldn't.*

"The amazing thing about welcoming dogs is how they fundamentally shift the energy," Brian explained. "When dogs are here, people just open up more and talk to each other. We've really tried to reach out to dog lovers, because they bring the energy we want."

Some five thousand dogs have come through Barrel Oak's doors since 2008. The winery specifically targets dog organizations and has hosted fundraisers for breed rescue groups. Brian told me about the day the winery hosted eighty Golden Retrievers and twenty Great Danes — at the same time.

"The owners of the two breeds were so different," he said, smiling as he recounted the story. "The Great Dane people" — he tensed his body and widened his eyes — "were all very austere, serious, kind of freaked-out personalities." To imitate the Golden owners, he waved his arms and wagged his tongue like a hyper ten-year-old.

Just then, Birch, who'd been balancing precariously with his front legs on a coffeetable and his back legs on the couch, tumbled over and crashed to the floor with a thud. He bounced up quickly, but not before a visitor joked, "How much has *he* had to drink?"

A decade ago, dogs in a tasting room would have been unthinkable. But an increasing number of new wineries, many of them targeting young professionals, have quite literally gone to the dogs. In California alone, there's Punk Dog, Dog House, Weener's Leap, and Mutt Lynch, which features whimsical illustrations of dogs on its wines, and names like "Unleashed" and "Canis Major."

Though there's always a risk of alienating wine enthusiasts who dislike dogs, Brian and others says it's one worth taking. "I can pretty confidently say that if we weren't as dog-friendly as we are, we'd have a failed business here," he said. "Yes, our wine speaks for itself. But you need to get people through the door. By reaching out to dog lovers, we've been able to do that."

Later that afternoon, as Casey and I prepared to hit the road again, Brian offered me one last glass. It would have been my third, and if I wasn't such a lightweight I might have accepted it.

"It's a kind gesture," I told him. "But I have enough trouble driving the RV when

I'm sober."

"Oh la vache!" I screamed a few minutes later on a quiet country road, as we drove past the first cows of our journey.

I'm unusually drawn to cows. During my summer travels around France as a kid, I would scream "Oh la vache!" every time we passed some in the countryside. Usually I'd demand that we stop to say hello. This proved impossible if we were in a train, but my mom would sometimes humor me if we were traveling by car.

"Oh la vache!" means "Oh the cow!," but the French use the expression to convey surprise. "Oh la vache!" essentially means "Holy shit!" As a ten-year-old, saying "Oh la vache!" allowed me to cuss without technically cussing. It was the one upside I could see at the time to my rigorous bilingual education.

Casey and I pulled over and crossed the two-lane road to get a better look at the herd. The black, brown, and white cows were busy eating from a trough in the shade at the bottom of a sloped field. The trough was only a few yards from the road, and as we approached a chipped orange stock gate that separated them from us, the smell of manure overpowered us. Or, I should say, overpowered me. It probably didn't bother Casey.

There's a great YouTube video of a boxer puppy in Newcastle, England, meeting a group of curious cows. They take turns sniffing the dog's face, and everyone involved seems delighted by the cross-species introduction. But that's not how it worked with Casey and these cows. When a curious black cow approached the gate, my dog pierced the silence with three rapid-fire barks. The herd stopped eating. An irritated cow mooed with displeasure.

Soon after, we came upon another first for our trip: a dog hanging out in the road. I saw him in plenty of time to slow the Chalet to a crawl. He was a black mutt, and I assumed he belonged to the owners of a farmhouse only a dozen yards removed from the road. The dog and I stared each other down for a few seconds before he bowed his head and shuffled toward the house. I wondered if this is how he spends his days — a kind of rural canine crossing guard. It didn't strike me as the safest of jobs.

Twenty minutes later Casey and I arrived in the village of Flint Hill, which is only two miles east of Shenandoah National Park. The Public House is easy to spot — it's a two-story white building perched on a small hill overlooking the village's main road. A former schoolhouse, it's now an upscale restaurant and hotel owned by William Waybourn and Craig Spaulding.

The expression "pet-friendly" gets tossed around liberally these days, but Craig and William take the promise seriously. Their restaurant's extensive "Pet Menu" includes a pooch stew and a grilled steak, and at the time of my visit dogs were allowed to dine at the feet of their humans in the hotel's upscale restaurant. So that's what we did. I don't think I've seen Casey happier than when the waiter brought him his "quarter hounder" at the same time as my strip steak. (The county health department has since put the kibosh on all that, though dogs can still get served on the patio.)

Perhaps as a reward for feeding him so well, after dinner Casey cuddled up to me on the room's king-size bed while I watched *Family Guy*. Brian Griffin, the animated show's talking dog, might be my favorite character on television — and he would have certainly fit in at the Public House. Brian is probably the most human member of the Griffin family; he's certainly the only one capable of true empathy. Maybe more than the human characters, Brian is multidimensional. He's the show's logical but troubled straight man, elevating the family's otherwise asinine discourse.

The dynamic between the Griffins — and much of the humor in *Family Guy* — depends on each member of the family demeaning the others. Brian is no exception. Though he's often the only voice of reason, the dog's more canine characteristics allow the human Griffins to bring him to their level. Brian may have attended Brown University, but he's also afraid of the vacuum cleaner and can't help rolling around in garbage.

The fact that Brian is a flawed, alcoholic dog might be the only thing that makes his relationship with the Griffins believable.

The next morning, Casey and I shared a hearty breakfast (eggs sunny-side up for me, scrambled for Casey) before setting off on a long day's drive to North Carolina, where we were to be joined by a twenty-four-year-old college student named Sam. He's the cousin of an ex-boyfriend of mine, and I'd invited him to come along as my research assistant — shoot some video, transcribe interviews — while we drove to Florida. I was looking forward to the human company.

I took the scenic route toward North

Carolina, down the 101-mile Skyline Drive through the Blue Ridge Mountains and Shenandoah National Park, home to one of the country's densest population of black bears. I'd spent many years living on the East Coast, but I didn't remember seeing a stretch of road like this one. There were some seventy overlooks, each more breathtaking than the next. From some you could see many miles into the valley below — from others, the Blue Ridge's smoky peaks and the Alleghenies.

The Appalachian Trail runs for 101 miles through Shenandoah, and it crosses Skyline Drive twenty-eight times. After stopping to eat a sandwich on a rock overlooking a river, I took Casey for a walk along a portion of the trail that hugs the mountainside. If we'd continued for 860 miles (or two million steps, as a sign informed us), we'd eventually come to the trail's southern end on Springer Mountain in Georgia. A blind man, Bill Irwin, once walked the nearly 2,200-mile trail with the help of his German Shepherd seeing-eye dog named Orient.

But though Casey and I didn't make it more than a mile or two into the forest, I'll never forget our walk on that clear, unseasonably warm February day in the shadow

of the Blue Ridge Mountains. I was happy.
Casey was deliriously happy. At one point I
stopped to tie my shoe, and he raced back
toward me with a thick stick in his mouth.
He nearly decapitated me with it, so I bear-
hugged him and we rolled around in the
dirt. I laughed. He barked. I turned over on
my back in the middle of the trail to catch
my breath. I closed my eyes. When I opened
them a few seconds later, I saw a red-tailed
hawk dancing above us in the sky.

4.

IN WHICH I DRIVE DOWN HILLBILLY CIRCLE, UP WIT'S END WAY, AND PAST HELL FOR CERTAIN ROAD

The more I drove around America, the more I yearned for an expedited RV park check-in process.

Was there not a line for motorhome travel

VIPs? Must I really be subjected to another round of mindless banter about weather and road conditions? Did I really need to listen to another recitation of a park's amenities in an RV park office with a wall of sewage hoses for sale?

The Birchwood RV Park in Durham, North Carolina, didn't bother with any of that. There were few amenities to highlight. There were no restrooms or showers. There was barely even a front office — it was a barren room with a telephone, a fax machine, and two milk crates for chairs. What Birchwood had instead was a short, eccentric Vietnamese manager named Pham.

Not long after Sam and I arrived, Pham came peeling around a corner in a beat-up minivan, a cigarette dangling from his lips. He skidded to a stop next to the Chalet and practically leaped out of the driver's seat to welcome us.

"How it hanging?" he said cheerfully, a turquoise polo shirt hugging his potbelly.

Eager to say hello, Casey ran toward Pham and nearly bowled him over. "What the hell!" Pham said, deftly rotating his hips to avoid impact. "Damn, dog, don't make me throw you on grill!"

Pham liked joking about tossing unruly dogs on grills. He liked joking about a lot of

things he probably shouldn't, but he buffered his inappropriateness with hospitality
and unexpected charm.

"I'm crazy motherfucka," he explained,
leaning against one of the park's many tall,
skinny birch trees. "I'm nice guy, but nobody messes with me. You never meet a
nicer guy than me, but when you make me
an asshole, you've never seen a bigger asshole."

Fortunately, Pham seemed to like Sam
and me. He assumed, I think, that we were
a couple, though I explained that Sam was
my research assistant — and that I was driving around the country for a book about
dogs. "Yeah, people crazy about dogs,"
Pham said, nodding in agreement with
himself. "We have a lot of dogs around here.
Lotta small, yappy ones. They're all right,
but they can get kind of annoying sometimes."

"Ever want to throw one on a grill?" I
asked.

Pham laughed. "Hell, yes! But I don't. I'm
nice guy."

He was indeed a nice guy. He patiently
explained to me how to fill the RV's water
tank (so we could have water on the road)
and how to attach a water hose to the
camper (so we could connect to a water

200

source at RV parks). I'd planned to wait until Florida to bother with this, but the lack of restrooms and showers at Birchwood meant we needed running water in the RV. And the weather was warming up; no need to worry about freezing pipes.

Though I didn't think it possible, Sam would turn out to be even more useless than I when it came to hoses, spigots, and most other practicalities of Chalet living. Whenever something mildly unpleasant had to be done — emptying the toilet's "black water tank," for example — he would disappear with Casey.

Pham quickly gave up trying to explain things to us. "I do it myself — you guys like *Two* Stooges," he told me as he tested the water pressure inside the motorhome. I was apparently missing a device designed to regulate water flow into the RV, and Pham hollered a series of expletives when water sprayed from the faucet and soaked his shirt. While he dried his belly with a paper towel, I tried to kill a mosquito with a rolled-up issue of *Vanity Fair.*

"What, you afraid of bugs?" Pham said.

"It's a *mosquito,*" I explained. "I won't be able to sleep with a mosquito buzzing around. I've managed to go three weeks without one in here."

201

"That's not mosquito. Looks like one, but it's not. Just bug."

"I think I know what a mosquito looks like," I insisted. "I have nightmares about mosquitoes."

He laughed. "How you gonna survive when you sleep outside?"

"I'm not sleeping outside — I'm sleeping *in the RV.*"

I swatted at the bug again and missed.

"You funny," Pham said. "You spend all this energy trying to kill a mosquito that not even mosquito!"

As I stalked the mystery bug through the motorhome, I couldn't help but notice what a stellar cleaning job I'd done earlier in the day. I'd spent an hour tidying up before Sam's arrival — I scrubbed the bathroom, folded my clothes, swept the floor for Casey hair, and made space in the cab for Sam to sleep. (He's even taller than I am, and I was relieved when he fit up there.)

I'd met Sam only twice before inviting him along for a week. He'd struck me as friendly and charming, but those who knew him better than I said he could also be moody and withdrawn. At twenty-four, he'd yet to marshal his deep intellect into conventional markings of success. He'd been in and out of college several times. Though I

202

couldn't relate to that specifically, I understood what Sam meant when he lamented being misunderstood, sometimes by his own design.

"My default mode can be pretty melancholy," he told me, "but I'm good at not showing it. I keep a pretty tight lid on myself most of the time; lots of people have described me as hard to know."

"Well, that makes two of us," I replied. "We'll be two hard-to-know guys driving around the country in an RV with a dog."

I told Sam that for long periods of my life I've felt like a phony. Though I crave true intimacy, the idea that I can reveal all of myself (warts and all) and still be loved sometimes strikes me as preposterous. For years, I rarely let friends, family, and lovers as close as they wanted to get. Because of that, any love they expressed felt fraudulent — worse yet, inaccurate. And though I wanted the love of those close to me, I didn't always value it; I was sometimes more concerned with what complete strangers thought of me. Did they perceive me as smart, stable, put together? Were they attracted to me? Was my shiny, phony facade holding up? Was I admired from a distance?

Though all of that has lessened with age (and therapy), I told Sam that I sometimes worry that Casey might be the only living creature that knows the truth about me.

On our first night at Birchwood, Sam announced that he would cook us dinner.

He shook his head in disbelief at the frozen pizzas and TGI Friday–brand microwavable fajitas in the RV's small freezer. "You shop like a fourteen-year-old whose parents left you $100 for groceries while they're out of town," he later told me.

That characterization wasn't entirely fair. I'd recently run out of iceberg lettuce, and there were carrots, yogurt, and several kinds of fancy cheeses in the fridge. But I did see his point. "We'll go shopping tomorrow," I assured him.

As we ate microwavable fajitas and drank wine that night, the motorhome's gas leak detector went off, blaring an ear-splitting warning that sent Casey scrambling for the Chalet's door. I'd accidentally spilled a bottle of ethanol-based cleaning solution next to the detector back in Virginia, and every few days the alarm would pierce the quiet.

After dinner, I emailed the media relations department of ThunderShirt, the makers of a dog antianxiety compression wrap of the same name. (It fits around the dog's midsection like a sweater.) A number of my Facebook followers had recommended it to me after the failure of the homeopathic "calm down" medicine. In my email, I explained Casey's predicament and asked if they might overnight a ThunderShirt to the RV Park. Should it work, I hinted, I might write about it in the book.

Two hours later, the company's CEO, Phil Blizzard, wrote back in the affirmative. "And if you're ever in the Durham, North Carolina, area, please stop by."

I couldn't believe my eyes. "Wow, I am actually in Durham right now," I wrote back. "What are the odds?"

The next day we drove to ThunderShirt's headquarters, located in what used to be a car dealership near the home field of the Durham Bulls, a Triple-A baseball team made famous by one of my favorite movies of all time, the 1988 sports classic *Bull Durham*.

In a conference room, Phil insisted that his own dog — a Golden/Poodle mix named Dosi — inspired his invention. Dosi, he told us, would shake and pant during thunderstorms and fireworks. "I would wake up in the middle of the night with her standing on my chest, vibrating and staring into my face."

Phil tried sedatives, but they're expensive and made Dosi a danger to herself. (Her legs would crumple under her when she jumped off the couch.) Phil played recordings of thunderstorms on his home stereo system — softly at first, then louder — hoping to desensitize Dosi to the scary sound, but she knew the difference between a fake storm and a real one. Finally, a friend of Phil's suggested he tie a T-shirt with tape around Dosi's torso during a storm, as a way of applying calming pressure.

"I'm an engineer by training, and I thought it was one of the stupidest ideas I'd ever heard," Phil told us.

But he was desperate enough to try it during the next storm, and to his surprise it worked. "It was like flipping a switch with her," Phil said. He launched ThunderShirt a year later, and the company has quickly grown into a pet industry success story.

According to a survey of American dog owners commissioned by ThunderShirt, 41 percent report having a dog with anxiety problems, most commonly due to loud noises (thunderstorms, fireworks) and separation anxiety. Dog trainer Cesar Millan, whom I would be visiting later in my journey, has famously said that America's neurotic and anxious dogs are a reflection of their neurotic and anxious owners. So, was I to blame for Casey's fear of fireworks and moving motorhomes? I wasn't convinced. Besides, it seemed like just another thing for me to feel guilty about.

"I'm really only focused on making Casey comfortable in the RV," I told Phil. "You can't imagine how stressful it is to drive around the country with a dog who hates driving around the country."

But I was skeptical that pressure was the solution to Casey's anxiety. Before I'd even heard of ThunderShirt, I would instinctively hold Casey tight during thunderstorms, hoping the close touch would calm him down. But it never did — it was like I wasn't even there. Phil said this didn't automatically mean Casey wouldn't respond to the ThunderShirt, but it lessened the chance.

"ThunderShirt doesn't work for every dog," he cautioned us. "We don't know why some dogs don't seem to respond to it."

"Maybe the nonresponders are the *smart* ones," I said, half-joking. "Maybe they're the ones who realize that a shirt isn't going to save them from fireworks, or a loud, moving motorhome."

Before we left, Phil gave us two free ThunderShirts — one gray, one blue — and wished us well. "Do you think this is going to work?" I asked Sam in the parking lot as we huddled over Casey and eventually figured out how to secure the ThunderShirt to his torso.

"I don't know," Sam said. "But he looks damn sharp in it."

Two days later, high in the hills of Maggie Valley, North Carolina, six Huskies led us up a steep climb to Hemphill Bald, a meadow that starts 5,500 feet above sea level and borders the Great Smoky Mountain National Park.

Like so much of Appalachia, Maggie Valley stuns visitors with its breathtaking beauty — and extreme poverty. When we'd rolled into town that morning, the landscape was littered with rusted tractors and slanted, skeletal timber houses and barns. The tourist traps along the town's main drag — kitschy shops, putt-putt courses, a solitary snow tubing store — were shuttered or barely hanging on. The zoo was empty. Even the area's iconic Wild West–themed amusement park, Ghost Town in the Sky, sat unused at the top of a mountain, a literal ghost town in the sky.

Maggie Valley was once the heart of moonshine country (famed moonshiner Marvin "Popcorn" Sutton grew up here), but these days meth labs have replaced some stillhouses. We drove by abandoned labs dotting the hillsides, their roofs charred black.

At the top of Hemphill Bald, though, it was easy to forget about all that. To the north and northwest stretched the Cataloochee Valley, an isolated area once favored by Cherokee hunters. To the south we could see the Plott Balsams, a mountain range with five summits of more than 6,000 feet. The state dog of North Carolina is the Plott Hound, a fearless and deeply revered hunting dog. In a glowing article about the breed, Richard Woodward called Plott Hounds "the ninja warriors of dogdom" because of their willingness to "grapple with a baying 500-pound bear eight times its size."

A few yards to my right, Casey rolled on his back in the sun-stained grass, his neck stretched like a goose's. Behind me, Sam kneeled next to the leader of the Husky pack we were climbing with, a headstrong female nicknamed Queen Natasha the Evil. She belongs to Kirk Wall and Todd Fulbright, who spend every free minute they have hiking in these mountains with their six "misbehaving" Huskies they call the Thundering Herd.

Dogs aren't allowed on most trails in the Great Smoky Mountains National Park, so Kirk and Todd instead took us through the hills of the adjoining Cataloochee Ranch. Casey loves hiking, but he gave us all a scare as we made it down from Hemphill Bald. Normally he's cautious around things he hasn't seen before, but he surprised me by doing the dog version of "Oh, to hell with it" and leaping straight onto a cattle guard. All four of his legs crashed through the rails, his knees banging against the steel. I gasped. He yelped. Then he panicked. He tried to lift himself up, only to fall through again. As I lifted Casey in my arms, his body stiffened — he was in shock. I carried him across the guard and placed him gently on the soft dirt. He hardly left my side for the rest of the hike.

The next day, we left Casey in the Chalet before meeting Kirk and Todd again. Casey was fine and had seemingly forgotten about his trauma, but he wasn't allowed where we were heading. "I can't believe we're meeting wolves today!" Sam said as Kirk, who wore a sweatshirt of a stenciled Husky covered with a coat of fluffy dark hair from Kirk's shedding dogs, drove us past Hell for Certain Road, and then up a narrow, rutted dirt road called Wolf Tail Way. Kirk continued up a steep hill, past boulders and trees that opened up to a pristine stone house and, behind it, a small cedar smokehouse.

We were greeted there by a short, silver-haired man built like a tree stump. His name is Rob Gudger, but some folks in these mountains call him Wolf Man. "Good thing you didn't bring your RV — you'da died before you got here," Rob said with a broad smile, peeling off a pair of thick, dirty work gloves to shake our hands. He spoke with a friendly mountain drawl, the kind that should narrate every story.

Though it's illegal to keep wolves as pets in this country, some states, including North Carolina, allow wolfdog hybrids. There are some 300,000 in the United States, far more than any other country. Rob has four — we could hear them yelping and howling from behind his house.

"For it to be legal in North Carolina," Rob explained, "you gotta call 'em hybrids." He said his are 97 percent wolf, 3 percent Husky.

"How often do they howl like that?" I asked, eager to see them.

"Well, you know, in the twenty-five years I've lived with wolves, I don't think I ever heard one *howl,*" he said. "They sure can *sing,* though. I try to explain to people that the difference between singing and howling is like the difference between bluegrass and opera. I'm an ol' flatfoot clogger who happens to like bluegrass, so I happen to think they sing. But you can think they do whatever you want."

He brushed some bits of bark from his heavy red flannel. "Thing is, it's easy to get 'em started singing. Not so easy to get 'em to stop." He chuckled. "What do y'all say we go meet 'em?"

As we rounded the house and approached their fenced-in enclosure, the wolfdogs paced, yelped, and head-butted each other in excitement. They were long-legged, beautiful animals. Two were dark gray. One was pure white. The fourth — the youngest — was brindle. We watched as she climbed up the pen's fence, hooking on with her claws and dragging herself up.

"She wants to say hello," Rob said.

Or eat us, I thought. I remembered reading somewhere that the ASPCA has called for an end to wolfdog breeding, which it considers dangerous and unpredictable.

"I didn't know wolves could climb," Sam said, mesmerized.

"Yup, they do that," Rob told us. "They're always trying to escape, so you just gotta stay a step ahead of 'em with wire and a little bit of electricity."

Rob picked up a shovel. "Now, don't y'all worry, I've never had to use this for what you might be thinking," he said, his breath visible in the crisp mountain air. "These guys won't hurt you. People see wolves and they stick their own misconceptions on 'em. Wolf'll try to be playing and they freak out and say, 'He tried to bite me!' Bullshit! A wolf don't *try* to bite you. If he bites you, yur' coming back to me with a stump. Wolves are shy, and misunderstood."

Rob encouraged us to put our hands against the fence. "They'll lick you to death," he said.

"Go try it," I whispered to Sam.

He shot me a look. "Oh, nice, use me as a test case."

Kirk and Todd didn't share our reluctance. They'd met these animals before, and they sauntered right up to the fence and gave up their hands. The wolfdogs shouldered each other out of the way to lick them.

"They ain't gonna bite you," Rob assured us.

Sam proved braver than I. And after watching the white wolfdog — the pack's Alpha female — take a liking to the back of his hand, I shoved my notebook in my coat pocket and surrendered my own. The two gray wolfdogs each went to work on one of my hands with warm, soft tongues.

"Okay, I'm gonna head in," Rob said a minute later, opening the first of two gated doors to the pen. The wolfdogs promptly lost interest in us and started play-biting each other, yelping with anticipation. The brindle wolfdog climbed up the fence closest to where Rob was making his way inside.

"They'll calm down in a minute," he assured us, though we didn't mind at all. "They're crazy right now 'cause they think they're getting fed, or going for a ride in the truck."

Rob occasionally transports the animals to one of his "Wolf Tales" educational events at churches, schools, or private events. All of his wolfdogs behave around people, he explained, but he usually showcases the gray ones. "That's what people expect a wolf to look like," he said.

"Wolf Tales" is part history lesson, part wolf myth busting. He tells his audiences that wolves in the wild are afraid of humans, and that they're significantly less dangerous than coyotes or dogs. But he's up against what he calls the fallacy of the Big Bad Wolf. As Brian Hare and Vanessa Woods argue in their book, *The Genius of Dogs,* "no other animal has been portrayed so ubiquitously as the Bad Guy throughout history."

It's surprising that wolves get such a bad rap. Without wolves, of course, there would be no dogs. The history of that evolution is still largely up for debate, but most current estimates place the start of domestication about fifteen thousand years ago in East Asia, while one recent study suggests a much earlier start date, with a change of location — as early as 32,000 years ago in Europe.

Rob's wolfdogs go to the vet several times a year, just like regular dogs, and each animal is named "Wolf" in a different Native American language.

"Do they know their names?" I asked.

"Nah, wolves don't care what you call 'em," he said. "The only reason I named 'em is 'cause the vet said he needed names for his records. You can call wolves by their names all day long, and they won't even look at you. You can't train these animals like you can a dog. Wolves don't want to please you."

Though one of Rob's neighbors keeps a wolfdog as a house pet, Rob doesn't recommend it. "For one thing, I'm a neat freak," he told us. "But these guys are just not meant to be inside. I respect 'em and they respect me, but they're not going to sleep in bed with me."

That echoed what I'd heard from Amanda Shaad, who has worked with pure wolves at Wolf Park in Indiana since 1995. "With a wolf, you're getting an animal that hunts for a living," she said. "With a dog, you're getting an animal that likes to be around people. So with a wolfdog, you've created an animal that's not afraid of people — and hunts for a living. Does this sound like a good idea? You can end up with an animal that likes hanging around humans but will guard the couch from you so they can disembowel it, or guard the refrigerator from you because food guarding is part of their behavior."

But Shaad conceded that it's difficult to generalize about wolfdogs, who can have wildly varying percentages of wolf and dog. "A lot depends on the breed of dog the wolf was bred with," she said.

Rob told us he'd never want to see a wolf bred with a German Shepherd, or a Malamute. "Mine have Husky in 'em, and you can depend on what Huskies are gonna do," he said. "They weren't bred to be aggressive or protective. If a wolfdog goes after a human, you can bet it's the dog part of 'em that's at fault."

Playful as puppies, the wolfdogs romped

around with Rob in the pen as he talked. We watched as they leaped toward his face and play-bit his gloved hands. When they got too rowdy, he would swat them away with his forearm.

"Are you ever worried they'll hurt you?" I asked.

Rob shook his head. "Nah — they know who feeds 'em," he said. "Besides, they just want to play. People get all frightened by 'em, but you gotta realize that they've got this big head full of teeth, and these big 'ol paws with big claws on 'em, and that's all the equipment they've got to play with. What hurts us don't hurt them, so they don't know any different. So as long as I'm not so stupid as to stick my face down at 'em, all they can do is chew me a little. Ain't so bad."

I wondered what drew Rob to these animals. He told us he considers himself a cross between a caretaker and a warden, doing his best to keep them alive, healthy, and far from the neighborhood cats. But why wolves? Wouldn't some run-of-the-mill dogs make better companions for a man who lives alone in the mountains? Though the wolfdogs were playful with Rob during our visit, he told us they aren't affectionate animals. And he rarely enters their pen, except to feed them.

"What does a wolf bring you that a dog doesn't?" I asked.

"Man, dogs have been dumbed down so much by humanity," he said. "These guys, they've still got all their senses and faculties about 'em. You watch 'em, and they're always sniffing, looking around, *watching.* And that's what I want from them. If I could ask one thing of 'em — and obviously I can't — I'd just ask them to act like a wolf."

Rob had told us earlier that wolves are shy and misunderstood. Did he also consider himself shy and misunderstood? Was that the connection?

He looked down at his rowdy pack. "Yup, that's probably why I like 'em so much. I feel like I've been misunderstood my whole life. Ain't nothing that drives me crazy more. It's the worst thing in the world to be shy and misunderstood. I've mostly gotten over the shy part, but I've never gotten over being misunderstood. I used to be a people pleaser — tried to get along with everyone, make everyone happy. But that's something even the Bible tells you that you don't gotta do."

It's also something wolves never do.

"We bred dogs to be needy," he went on.

"Wolves are independent. They're smart as hell. But they can't talk, and they need someone to speak for them, to set the record straight. They get blamed for a lot o' things they shouldn't get blamed for. So that's why I do what I do, trying to educate people. That's why I save every last penny I got to feed these guys, keep 'em healthy and safe in their pack, and take 'em out and show the world that they don't mean no harm to anyone."

"Why don't I be the DJ for a bit?" Sam said as we departed Maggie Valley the next day for the 335-mile drive south to Savannah, Georgia.

The poor guy had suffered through a few days of my iPhone library favorites — REM, Moby, Florence + the Machine — and was eager to broaden my musical horizons. First he made me listen to Manu Chao, an eccentric French rocker of Spanish descent who sings in multiple languages, sometimes in the same song. Then came Childish Gambino, an American rapper Sam respects for being clever — and more "emotionally honest than most rappers." (In a song titled "Bonfire," Gambino utters this line: "I love pussy. I love bitches. Man, I should be running PETA.")

While we were on the subject of rap, I had Sam listen to "Saving Seamus Ryan," an album by a dog-loving Boston rapper named Esoteric. Though many rappers like to call themselves some variation of "dog" (there's Snoop Dogg, Top Dog, Tim Dog, Phife Dawg, Tha Dogg Pound, and so on), Esoteric likes rapping *about* dogs.

"Saving Seamus Ryan" is Esoteric's goodbye to his dying Lab. One of the album's best songs is called "Back to the Lab."

women come and go but a dog stays
    always
all days, hanging by the couch or the
    hallways
the simple sound of me jingling my keys
will generate a look that could bring me to
    my knees
he's the type of friend that you can't stay
    mad around
even when you hear that shatter sound, it
    don't matter now

it's all good, i got a lab, the love's essential
i saw myself as garbage he saw the
    potential
and when i can't grin or lift my chin
there's nobody, more happy to see me than
    him

yes i'll never turn my "back to the lab"
and i'm always goin' "back to the lab"

We blasted Esoteric's music as we drove toward Asheville, where we were going to have lunch. It was Casey's third day wearing the ThunderShirt, to little effect. Sam and I would occasionally try to convince ourselves that we saw improvement (*Didn't Casey just seem to relax there for a second?*), but he was as uncomfortable as ever. He would stand up, sit down, stand up, sit down. Sometimes, he would sigh loudly and plop his snout in Sam's lap.

As we lumbered through the outskirts of Waynesville, North Carolina, thirty miles west of Asheville, we came upon a road sign advertising a bookstore with more than 200,000 books. "We have to check it out," I said, following the sign toward a two-lane country road. We drove for several minutes — up a hill, past some cows — before I was pretty sure we were lost.

"I think we missed it," I mumbled.

"Let's give it another mile," Sam said.

Just as we were about to give up, we saw a small, stand-alone blue block structure with a sign next to the front door that read *BooK Store.* We had to navigate a short, steep decline to reach the building. The Chalet's nose dipped precariously; for a moment, we could see only concrete through the windshield. Sam's body tensed as my laptop slid off the dinette bench, landing with a muffled thud on Casey's bed.

We survived the descent and parked in the small empty lot, next to a battered basketball hoop with the net tangled around its orange rim. Across a footbridge was an old redbrick house. A dog barked in the distance. "This place looks too small to have 200,000 books," I said as we approached the store's front door, where a handmade sign beckoned us to *Come In.* As soon as we did, our jaws dropped.

This wasn't so much a bookstore as an episode of *Hoarders.* Everywhere we turned, we saw books piled floor to ceiling. Stacks that didn't reach the ceiling leaned precariously into the aisles; shorter piles jutted out at our shins, nearly blocking access to certain rows. In the rear of the store, stapled to a shelf on which rested an old copy of *All The President's Men,* a yellow paper with black ink read, *Go on steal that book. You will not be happy when you are in hell. God is watching you.*

I heard the door swing open. "What you boys looking for?" said a woman with a mountain twang that made "Wolf Man"

Rob Gudger sound like a city slicker.

I shimmied my way back toward the front. "We saw the sign that said 200,000 books," I said, stepping over a pile of romance novels. "We had to see this place."

"Oh, that's an old sign," she said, flicking on the store's lights. "There's probably more now."

I turned a corner and got my first look at Mary Judith Messer, who has owned the rural bookstore — and lived in the house next to it — for forty years. A youthful seventy-four, she had big blue eyes and straight, boyish brown hair with uneven bangs that partially covered her thick eyebrows. She wore jeans and a puffy blue turtleneck, and a long silver cross necklace rested against her midsection.

Mary seemed delighted to have company. "Religion and classics are down there to your right; science fiction and fantasy is over there to the left," she said, pointing enthusiastically in various directions. "This here on the left is war. This is mystery. This here is biography." She led me a few paces. "Back on that through there is history. Health is down that aisle."

"Do you have a pets section?" I asked.

"Of course!" she said, pointing toward a dozen shelves overflowing with titles like *A*

Guide to Owning Goldfish, The Ultimate Cat Book, and many how-to books about dogs — *How to Raise and Train a Siberian Husky, How to Talk to Your Dog, How to Be Your Dog's Best Friend* (by the Monks of New Skete).

When I explained that I was driving around the country to write a book about dogs, her eyes widened. "I wrote a book, too!" she said, handing me a paperback copy of her autobiography, *Moonshiner's Daughter: Growing Up Poor in the Smokies . . . How Did We Survive?*

While Sam wandered through the store, I skimmed the first few pages of her book. They were heartbreaking. On page 6, Mary is molested by a school janitor. On the next page, she's beaten by a teacher. A few pages later, Mary's father — an alcoholic moonshiner — pummels her mentally ill mother with a piece of stove wood in the small, dilapidated shack where they lived. Even the family dog (an old mutt named Brownie) has it rough. He barely survives being bitten under the eye by a copperhead, and one winter day he just wanders off into the mountains.

"Just as well, I guess," Mary writes. "Most people fed their dogs table scraps and we never, ever had any. I guess he took up with

another family who did." In reality, he probably left — as some dogs do — to die alone.

"Do you have a dog today?" I asked her.

"Oh, yes," she said, her face beaming. "A
little Chihuahua named Tootsie Roll, and
an outdoor dog named Bear."

Mary told us that except for the decade
she spent in New York City, she's always
had dogs. "People with animals live longer.
Especially dogs." She stood behind her cluttered front desk as she spoke, wiping dust
off books with a rag. "I'm not no *cat person*.
My mother was a cat person. It's not like I
dislike my mother; it's just that I never really
hooked up with cats. I'll let 'em be, though.
They can live or whatever, you know." She
laughed. "But a dog! I mean, they're just
everything to me."

"How so?" I asked, before offering the
question I would ask over and over again
during my months on the road. "Why are
dogs so meaningful to you?"

"Well," she said, "I guess it's because
they're just always there — you can count
on 'em. You can make a mistake or not treat
'em as good as you should sometimes, and
they'll still come back and love ya. I think
God put dogs on this earth so that we would
take care of 'em, and they would give us
companionship. Think about the lonely old

man or woman — every human that's important to 'em might be gone, but they still get companionship if they have a dog. The dog's the faithful friend that lies by their chair when they watch TV or read the paper. If they walk to the porch, the dog's at their side, looking up at 'em like they're the most important thing in the world. Which they are!"

I asked Mary about Bear, her fifteen-year-old outdoor dog. She told us that because of his unusual energy level ("He's the most hyper dog in the world") and his eagerness to urinate indoors ("He'll pee right on the TV"), that he's not allowed in the house. But he's also not allowed to roam freely, for fear that he might run onto the road.

Like many "outdoor dogs" in America, particularly those in poor areas, Bear spends most of his life chained up. Which, of course, means he doesn't get enough exercise — only adding to his perceived hyperness. Mary feels guilty about Bear's circumstances. She believes that if a dog can't be a close companion to a human, he should at least be allowed to run loose in the country.

"We're never going to have an outdoor dog again," she said. "We do our best to keep him comfortable — he has a long chain and has a nice dog house with pine needles so he won't get mange. I go out there every day and feed 'im and check his water. But we know good and well it's not the way it should be."

Bear probably needs a fenced-in yard, or an electric fence. "Sometimes I think that if someone had a big yard, that Bear could just run around and do his thing," she told us. "But my son doesn't want to give him up."

Sam and I could have spent all day in Mary's bookstore, but we had to get back on the road. Before leaving, I bought a copy of *Moonshiner's Daughter* and promised to send Mary a copy of *Travels with Casey*. She wouldn't let us leave without a fight, though.

"You know," she said, "I am going to have a movie made of my book, and with the money I will have a new store, probably over there" — she pointed behind us, toward the south — "in that great big field by my house. I want to make it four times as big as this. It's going to be huge. And *everything* will have its place."

She touched the cross on her chest. "I just hope the Lord don't wait until I'm too old to reorganize!"

The six-hour drive to Savannah felt like ten. We took Interstate 26 southeast through

South Carolina, until we connected with Interstate 95. Sam napped for part of the ride. I was tempted to wake him up so I'd have someone to talk to (or, better yet, so he could take the wheel), but he looked so comfortable curled up in the passenger seat. Besides, the RV insurance mandated that I be the motorhome's only driver.

We pulled into Savannah's Red Gate Campground & RV Resort as the sun was setting. The park used to be a Confederate soldier encampment. Today, its grounds are impossibly picturesque. There are grassy fields, red barns, whitewashed fences, oaks hung with Spanish moss, and two lovely ponds patrolled by herons. The temperature hovered in the upper sixties when we pulled in, but after weeks in the cold it felt like eighty. I changed into a tank top and shorts, grabbed a tennis ball, and sprinted toward the closest pond, with Casey trailing close behind.

"Wanna go for a swim?" I said, eager to see him in water for the first time in weeks. But just as I was about to hurl the ball into the pond, I spotted a small sign attached to a thick oak tree: BEWARE! SNAKES & ALLIGATORS.

Later that night, Casey and I took a walk through the park. Everything was swathed in thick fog. Moss-draped trees loomed out from the darkness, and a halo of warm yellow light hovered around the pavilion and clubhouse. Even the roar of a passing Amtrak train sounded romantic. I kneeled down next to Casey, grabbed him by the scruff, and reminded him that he was the "best boy in the whole world."

Before bed, I signed on to Twitter to check on one of my favorite accounts, @dogsdoingthings, which reads as if David Foster Wallace and Hunter S. Thompson got together in the afterlife to inspire acid flashbacks in the literary-minded. The account's creators, Clayton Lamar and Patrick Carr, delight in imagining dogs presiding over a dying earth — or something like that.

In my few weeks on the road, @dogsdoingthings had tweeted dozens of gems. These were three of my favorites:

Dogs holding out a map showing the whole history of your desire and casually tossing it at your feet, sneering "disgusting."

Dogs lamenting as the outer edge of history passes from tragedy into farce, "I was just — learning to — love."

Dogs dislodging a katana from the neck of a shuddering body and collapsing into an easy chair, sighing, "HONEY, YOU JUST WEAR ME OUT."

The next morning, Sam cooked eggs and bacon before we headed into town to meet Savannah Dan, the city's most famous tour guide. He gives two dog-friendly walking tours each day, wearing a cream seersucker suit with a bow tie and pocket square, white buck shoes, and a straw Panama hat. At six-five and 275 pounds, he's a sight to behold.

The tour began at Johnson Square, the city's oldest. Nathanael Greene, a trusted general of George Washington during the Revolutionary War, is interred under an obelisk in the center of the square. Nearby is a bronze sundial, a fitting touch considering Greene died of sunstroke.

"What kind of dog is Casey?" Savannah Dan asked a few minutes into the tour, which would take us through six of the city's public squares, most shaded by oak trees and adorned with a statue of a prominent local figure. There were several other people on the tour, but Casey was the only dog.

I told Savannah Dan that I didn't know for sure, but that I thought Casey was half-Lab, half–Golden Retriever.

"That brushy tail looks like a Blue Heeler's tail," he said. "You familiar with a Blue Heeler?"

I nodded.

"Australian Cattle Dog — great dogs. I have one at home, and he has the exact same brushy tail. Casey's also got that barrel chest, but it tapers off at the hips. That's very Blue Heeler–ish. So are the different color guard hairs on his back."

It was the first time anyone had compared Casey to a Blue Heeler, and, frankly, I didn't see the resemblance. But Savannah Dan spoke with such a deep, syrupy voice, that I didn't dare interrupt him. He was a fountain of historical knowledge and salacious city gossip. His stories took us back all the way to the founding of the Georgia Colony and its roots as a buffer between Spanish-controlled Florida and the profitable Carolina colonies, and as a destination for the dregs of the British poorhouses.

I never knew that Georgia was the only British colony to originally ban slavery, but Savannah Dan filled us in. "The rule back then was no Catholics, no Jews, no lawyers, no liquor, and no slaves," he said.

The no-slave rule didn't last long, of course, and Savannah Dan tried his best to find the upside of losing the Civil War. "So the North may have won," he told us, "but we ended up with NASCAR, sweet tea, and Lynyrd Skynyrd." He smiled. "It all worked out in the end."

As we strolled through Savannah's cobblestone streets, I thought of William Simon Glover, the elderly African American who used to walk this same city wearing a suit and a fedora and leading an invisible dog named Patrick. As the story goes (told by John Berendt in *Midnight in the Garden of Good and Evil*), William worked for years as a porter for the Savannah law firm of Bouhan, Williams & Levy. While Bouhan was alive, William served as Patrick's caretaker; he would walk the dog and give him Chivas Regal to drink.

Before his death, Bouhan stipulated that William receive $10 a week to care for Patrick. When the dog died, William went to court to say the payments should stop. The judge, though, wouldn't hear of it.

"What do you mean Patrick is dead?" he told William. "How could he be? I see him right there! Right there on the carpet."

It took William a second to figure out what the judge was up to. "Oh, I think I see him too, judge!"

"Good," the judge replied. "So you just

232

keep walking him, and we'll keep paying you."

When Berendt met William, he was eighty-five and had been walking a dead dog for twenty years. William would stroll through downtown each morning and late afternoon, and as he did townspeople would ask, "Still walking that dog?" William never broke character. He'd look over his shoulder and say, "Come, on, Patrick," to the invisible dog lagging behind.

If I wasn't going to bump into William on this walking tour (and I wasn't — he's dead), I was at least going to sit on the bench where Forrest Gump waited for the 39 Bus with a box of chocolates for Jenny. I knew this made me the worst kind of Savannah tourist, but I didn't care. "Are we there yet?" I badgered Savannah Dan, who wisely saves the Forrest Gump bit for the end of his tour.

You can imagine my disappointment, then, when Savannah Dan pointed to the north end of Chippewa Square. There was no bench. There had, in fact, never been a bench. Savannah Dan explained that Tom Hanks had instead relaxed on a fiberglass prop. It had been trucked in and placed at the edge of the park, facing out toward a one-way street.

"We didn't even know what movie they were making," Savannah Dan recalled. "All we knew for sure was that Tom Hanks looked like a moron, he was sitting on a bench that had never before existed in Savannah, and traffic was going the *wrong direction* around Chippewa Square."

At the end of the ninety-minute tour, I took a picture of Casey with Savannah Dan and asked him for a dog-friendly place to eat lunch. He pointed us toward J. Christopher's, a restaurant with a low-sodium dog menu.

"My dog would never eat there, though," he cautioned us. "He's an *American* dog, and he'd rather have the sodium. If it ain't salted, he ain't interested."

We took off toward Florida the next morning, but Sam and I decided to make a slight detour, to Waycross, Georgia, a small city in the southeastern part of the state. Waycross is home to the Southern Forest World Museum, which displays a mummified hound dog that was trapped in the hollow trunk of a tree.

We arrived at 11:30 A.M., but the doors to its log building were locked. A sign read, "Attention: Due to the economy, Southern Forest World is open now at greatly reduced hours — Tuesday–Friday: 11:45–3:30." A few minutes later, a ponytailed man in his fifties showed up, unlocked the front door, and led us through this small museum devoted to Southern forestry. We saw a timber-hauling steam engine from the early 1900s, a loblobby pine replica tree to climb in, and a NASA space suit composed partly of pine. But the main attraction was Stuckie, which the museum confidently calls "the world's most famous petrified dog."

A logging contractor discovered the brown and white dog in the early 1980s, about twenty feet up a chestnut oak tree, and only a few feet from an exit hole. It was estimated that the dog had been dead for some twenty years, but that a "chimney effect" resulted in an upward air current, causing the scent of the dead dog to be carried away and protecting the corpse from insects.

In 2002, the museum held a "Name That Mummified Dog Contest." CNN picked up the story, foot traffic in the museum tripled, and naming ideas — King Mutt, Woody, Stay, Pupenkomen — poured in from all over the country. (One person suggested the museum not name the dog, "since he couldn't come anyway!") But after two museum board meetings devoted entirely to the issue, the name "Stuckie," well, stuck.

Wedged in his tree behind two sheets of plexiglass, Stuckie manages to look both terrified and terrifying. Desiccation has pulled his lips back, giving him a leathery snarl. But there's also an undeniable panic in his eye — he's fighting for his life. It's clear that he tried valiantly to escape from the hollowed trunk, only to get stuck tantalizingly close to freedom.

I felt icky staring at the poor dog. So did Sam.

"Let's get out of here," he said.

We piled back into the Chalet and hightailed it out of town.

5.
IN WHICH FLORIDA ISN'T NEARLY AS AWFUL AS I EXPECTED IT TO BE

"Welcome to the jutting extension of America's groin," I told Sam as we crossed into Florida on I-95.

I can't take credit for that line; writer Joe

Friedman penned it in his essay "Why I Hate Florida." Friedman's takedown of the Sunshine State is seething with anger and jacked up on metaphor. "With her makeup off, her golden wig in curlers, and her teeth placed in a jar, we see the truth," he writes, comparing the country's fourth most populous state to an elderly prostitute.

I wouldn't go nearly that far, but it is nonetheless difficult to appreciate Florida if you grew up in California, as I did. To me, entire swaths of Florida manage to feel both ostentatious *and* drab. It's worth noting that on John Steinbeck's journey around America, he skipped Florida entirely. *Nothing to see here,* he seemed to be saying. I couldn't skip Florida (too many dogs to meet!), nor did I want to. After weeks of driving and interviews, I was eager to park myself on soft, milk-white sand.

But before I could do that, Sam, Casey, and I had to pay a visit to the Orange County Convention Center in Orlando, home to the world's largest annual pet industry trade show, Global Pet Expo. The massive exhibit hall included 2,487 booths, hawking everything from canine couture to dog treadmills to "paw-litical" car magnets (*Bark Obama, Mutt Romney, Ron Pawl*). There was no way to see everything, so I'd asked pet industry publicist Stephanie Krol to show us around. Her company, Matrix Partners, claims to represent "the bark of the town."

These are good times to be in the pet business. Americans spent $53 billion on pet products and services in 2012, up from $17 billion in 1994. The pet industry has proven to be largely recession-proof — it was one of only a few to grow during the downturn of 2008 — and has attracted some unexpected names: Old Navy now makes dog sweaters, Paul Mitchell offers high-end pet grooming products, Omaha Steaks sells dog treats, and Harley-Davidson boasts "Bad to the Bone" dog bandannas.

Our first booth stop at the expo was Stella & Chewy's, makers of raw frozen and freeze-dried dog food. Raw food "is really hot right now," Stephanie assured us, but I didn't need any convincing. I'd read countless articles about the trend, including a *New York Times* piece in which Bravo Raw Diet cofounder Bette Schubert claimed dogs are healthier when they eat like their "wild ancestors."

While some vets are skeptical of the diet, I'd heard great things about Stella & Chewy's and was delighted when the company's founder, Marie Moody, insisted that we take bags of freeze-dried raw patties for the road.

"Does Casey like beef?" she asked me.

"He's a *Lab,*" I reminded her. "He likes everything."

I suggested we rip open a bag right there and give Casey a taste, but when I spun around he was gone. So was Sam. I spotted them several booths down, near two gigantic sleeping Great Pyrenees. "Looks like you might need a leash for Sam," one of Marie's employees said with a wink.

Next we were off to the Company of Animals, a British-based firm that sells everything from training leads and collars to Nina Ottosson's toys for dogs. I'd never heard of Ottosson, but apparently she was a big deal. "She's a rock star of this industry," Stephanie explained, adding that Ottosson rarely leaves her native Sweden but made a special trip to Florida for the expo.

Through a heavy accent, Ottosson told us she began making toys twenty years ago after worrying that her pets were bored. She spent much of the next decade driving around Scandinavia, selling her "brain games" — many of which require a dog to lift a block or turn a disk to find a hidden treat — and promoting the idea that canines, like humans, need mental stimulation.

"Now there are many other companies," making these kinds of games, "but I was the pioneer," she said proudly, pronouncing it *peeohneer.*

Poor Casey couldn't reach the treats hidden inside Ottosson's Dog Twister, a particularly complicated circular toy that requires dislodging plastic bones and sliding paw-printed plastic covers to find the reward. Casey knew where the treats were, but he couldn't master either step and finally resorted to trying to bite off the plastic covers.

"This is so difficult, not even *humans* can figure it out," Ottosson said with a smile.

In *One Nation Under Dog,* Michael Schaffer writes about the "three-sided conflict" — anthropomorphism, atavism, and solipsism — at the core of the contemporary pet industry. As Schaffer explains it, the tension is between "the desire to treat pets as humans; versus the interest in allowing them to live as close as possible to what we imagine to be their natural state; versus the less altruistic inclination to have the whole experience be easy for our human selves."

Rare is the pet product that encompasses all three concepts, but Ottosson's toys come close. Her games are convenient for us humans — they allow our dogs to entertain themselves when we're busy. They clearly fulfill our desire to treat dogs as human children. And while dogs wouldn't have had access to man-made toys in their "natural state," Ottosson's games can be said to combat the dumbing-down of the domesticated pet. They're an attempt to make dogs use their brains and solve problems on their own.

As Stephanie shepherded us across the carpeted convention floor, I asked her about the latest trends in the pet industry. "Eco-friendly stuff, for sure," she said, pointing to the bamboo ball launcher I gripped in my right hand. I'd gotten it earlier that day at the booth for Chuckit!, a company that claims to be "revolutionizing the game of fetch."

It should come as no surprise that young dog owners — many of them health-conscious — expect "green" pet products. "Nearly any trend in human consumerism will soon appear in the animal market," Schaffer points out. That might explain companies like FurBulous Dog, which produces organic shampoo "specially made for your sensitive dog," or Yoghund, which makes "organic and all-natural" frozen yogurt.

With half of America's dogs considered overweight or obese, the pet industry understandably focuses on low-calorie treats and meals. PetSafe, a company based in Tennessee, was promoting one of its newest products at the expo: a liquid dog treat called Lickety Stik.

"It's great for training, because it's low-calorie and won't fill the dog up," a PetSafe representative told us at his booth. As he spoke, he fed Casey chicken-flavored fluid from what looked like a glue stick. Casey couldn't get enough of the stuff and practically French-kissed the roller-top ball that dispensed the liquid.

"It's like crack for dogs," Sam marveled.

"You're not the first person to call it that," the man said. "But it's *healthy* crack."

Stephanie had to leave us eventually, so Sam, Casey, and I spent the rest of the afternoon wandering the convention floor on our own. As we did, we took a moment to consider the most ridiculous pet invention of all time: the canine treadmill.

"I guess it's for people who are too lazy to walk their dogs?" Sam wondered aloud.

Makers of canine treadmills insist the

devices combat this country's canine obesity epidemic. They certainly do little to combat this country's *human* obesity epidemic (for some dog owners, walking their pet is the only exercise they get), but Sam imagined a scenario I hadn't considered.

"I bet there are people who have two treadmills next to each other — one for the dog, and one for them," he said. "They can get in shape together, all without having to brave the elements — or interact with any other real people or dogs."

I nodded. "It's the perfect solution for a country obsessed with privacy, liberty, and dogs. There's no better place than America for the guy who wants to be left the hell alone with his dog."

"You sure you aren't one of those guys?" Sam asked with a smirk.

There weren't many dogs in attendance at Global Pet Expo, which might explain why people would stop us every few minutes to pet Casey — and, if he was lucky, give him a treat.

Casey had never eaten so well. He sampled a plethora of delicious treats, including chicken-cordon-bleu-flavored dog biscuits from Bocce's Bakery in New York City. We also dropped by the booth of Halo Purely for Pets, which had generously donated enough food to feed Casey throughout our journey. Halo is co-owned by Ellen Degeneres, on whom I have a celebrity crush. I'd irrationally hoped she'd fly to Orlando to meet us.

By the end of the day, Sam could barely walk under the weight of everything we'd accumulated. We had food, snacks, chew toys, harnesses, glow-in-the-dark dog collars, a doggie bed, and a Frosty Bowlz dish designed to keep a dog's drinking water cool. I wasn't convinced Casey cared how cold his water is, but I took one for the road just in case.

"I never knew that half the things I'm carrying even existed," Sam said. "I mean, is all of this stuff really necessary?"

It was a good question, one *Time* asked back in 1974 when it argued that Americans were spending too much on their dogs and cats: "The U.S. pet set gets not only more nutritious meals but also better medical care and vastly more affection than the great majority of the world's people."

In a 1969 *New York Times* article about the pet industry, reporter Alexander Hammer noted (under the heading "Snob Appeal") that pet ownership seemed to be changing in significant ways. Americans were both increasingly "attributing human qualities" to their animals and expecting the pet industry to make living with pets easier. Hammer quoted several pet industry and advertising professionals, with the consensus being that a combination of anthropomorphism and human loneliness was fueling the pet industry's growth.

"The pet owner attributes the human qualities of taste, including eye, nose, and palate, to his dog," Hammer wrote. Later in the story, he quoted an advertising man who delved deep into the psyche of the American pet owner: "People desperately want to love and be loved and the care and feeding of these pets is really an extension of the human need to fulfill a void."

We know today that anthropomorphism and human loneliness are linked. A 2008 University of Chicago study found that the more socially isolated people feel, the more they will attribute human motivations and mental states to their pets. Studies also tell us that Americans feel more disconnected than ever.

"It appears that the more isolated we become as a society, the more pets we want — and the more we treat them like humans," James Serpell, the director of the Center for the Interaction of Animals and Society at the University of Pennsylvania, told me at a conference on dog welfare. "If you plot the growth of the pet population against the decline in traditional forms of social support from friends and relatives, the two lines cross each other rather remarkably."

In an essay published in *The Domestic Dog,* Serpell writes that "the dog has been granted temporary personhood in return for its unfailing companionship." But it might be more accurate to say that we've granted our dogs *babyhood.* We push them in strollers, keep track of them on pet-monitoring phone apps, enroll them in school (where, theoretically, they learn how to behave), obsess about the consistency of their bowel movements, and consider medicating them when they "act out."

"Our expectations are really going up," pet industry analyst David Lummis told *The New York Times Magazine* in 2008. "Owners want their pets to be more like little, well-behaved children."

The problem, of course, is that dogs are not children. As Serpell puts it, we humans are often disappointed when a dog "reveals too much of its animal nature." It helps itself to trash from the garbage can, pees on the couch, barks at the neighbors, and sometimes *bites* the neighbors. (Some 4.5

million people in this country are bitten by dogs each year.)

Pet industry professionals know that people are significantly more likely to rid themselves of a troublesome pet than a troublesome child, so they're doing their best to convince us that dogs are "four-legged little children," Serpell told me, "with desires, needs, and rights. And they're succeeding."

The day after the expo was Sam's last with me in the Chalet, and I was feeling sad. "I'm going to miss you, man," I told him when he joined me briefly by the pool at the Orlando Lake Whippoorwill KOA.

As I bronzed, senior citizens and their grandchildren frolicked in the water. Word around this KOA was that the pool's temperature gauge was broken, and a concave-chested man in his eighties scrunched his face in displeasure when he dipped his hairy toes into the water.

"I need some soap with this bathwater!" he said loudly, startling a young girl dangling her feet at the pool's edge.

Apparently these seniors were professional RVers, because they spent the next few minutes trying to remember where on their travels they'd encountered water this warm.

"Doesn't this pool remind you of that place in New Mexico?"

"What place?"

"You know. Back in 2009."

"You think I remember 2009?"

"Didn't you get sick out there?"

"Where?"

"New Mexico."

"That pool water was even hotter than this."

"Wasn't that in Arizona?"

"He gets sick *everywhere.*"

"I miss Arizona. Why the hell are we in Florida?"

"Don't swear in front of the little ones!"

"Hell is *not* a swear word! It's a real place where bad kids go. They should know their afterlife options."

After an hour by the pool, we headed north toward the Daytona Beach Airport. As we rolled north on 95, we passed a man and woman on a red touring motorcycle. "I think she's holding a *dog,*" Sam said in disbelief.

I slowed down to see the woman (who, thankfully, wasn't driving) cradling a little brown dog against her chest. The animal was wrapped in either a towel or a thin white blanket — we couldn't tell for sure. But we could see the dog's head poking out

249

from the top, its ears flapping in the wind.

I'd spotted countless traveling dogs in my weeks on the road. In southern Virginia, I'd been passed by a dirt-smeared white pickup truck with the biggest Confederate flag I'd ever seen strapped to its side. At first the flag blocked my view of the chocolate Lab standing precariously next to machinery in the rear bed. The poor dog didn't dare move, lest he lose his balance and go tumbling onto the highway.

My other sightings were less dramatic. Near D.C., I'd waited for a green light next to an old Peugeot driven by a heavily made-up, cigarette-smoking woman with a French Bulldog on her lap. North of New York City, I saw three Golden Retrievers in a Honda Accord, each with their head out one window. But the cutest sighting of all had happened the previous day near Orlando, when Sam and I spied a young girl holding her stuffed animal dog high out the back window of her family's SUV. Her imaginary pup needed some air.

When we arrived in Daytona, Sam and I drove past the city's iconic racetrack before pulling up to the airport terminal. As the motorhome came to a stop, Sam turned in his seat toward Casey and gave him a pep talk. "You're going to learn to love this RV!"

he said, grabbing my dog by his scruff.

On the curb, I gave Sam a hug and thanked him for his company. "Thank *you*," he said, smiling broadly. "This is the most fun I've had in a long time."

When Sam had disappeared through the airport's automatic doors, I climbed back into the Chalet, took my seat, and looked into Casey's big brown eyes.

"It's just you and me for a while, buddy," I said.

Casey slumped into his usual spot and sighed.

"Wanna go to the beach!" I proposed to Casey the next morning on a blue-sky day in Jacksonville.

We'd spent the night at a campsite under the canopy of tall trees in Kathryn Abbey Hanna Park, which abuts the Atlantic Ocean on the northeast side of the city. We'd made friends with some vacationing Canadians, who, like me, were driving around Florida in a motorhome in early March with their dog.

It was a warm and breezy morning, and I couldn't get to the beach fast enough. I practically skipped my way there, with Casey following close behind. We had a stretch of coastline mostly to ourselves, so we played fetch along the ocean's edge and then wrestled in the sand. When I'd worn Casey out, I spread out my towel and dozed off.

An hour later, Suzi Teitelman showed up on the beach as planned. A yoga and fitness instructor, she takes the Downward Dog pose more literally than most. Suzi is credited with inventing doga — yoga with dogs. (The shared "O-G" of dog and yoga practically demands a clever portmanteau.)

Though I was raised in San Francisco by New Agey parents, I had somehow never tried yoga, let alone yoga with a dog. But Suzi — a tan, energetic, dog-obsessed transplant from New York City — urged me not to overthink my beginner status. "Doga is really just about connecting and having fun with your animal!" she said as her three dogs — Tucker, a Cockapoo, Coali, an American Cocker Spaniel, and Curli, a Malti-Poo — romped in the sand with Casey.

"Are we ready to do some doga?" Suzi said a few minutes later, sitting cross-legged on a thin purple blanket. She wore a pink tank top, black spandex shorts, and big sunglasses that covered half her face. As she smiled at Casey, he barked at me to throw him the tennis ball.

"Are you going to be a good monkey?" she asked him in a baby voice. "You going to lay down and do some doga with daddy?"

To my surprise, Casey complied. He didn't even sigh as he spread out on the beach next to me.

"Good," Suzi said before taking a deep breath. "Now, we just start by sitting and meditating and massaging our dogs." She worked her hands softly into Coali's side. Tucker, meanwhile, meandered down the beach in search of something. "He's not really a doga dog," Suzie explained. "He doesn't like it as much as Coali does."

We sat quietly for a minute with our animals. "As we get calm and breathe gently through our nose, they get calm, too," Suzi said. She was right — Casey became so relaxed that he practically fell asleep. She then guided me through a handful of poses, some of which involved Casey. In one position, she had me sit and lean my back against his. "We're aligning our chakras," Suzi said. Next we did Child's Pose, which works the spine, hips, and thighs. (We petted our dogs as we stretched forward on our knees.) Then we did the Downward Dog — I leaned over Casey to form something approximating an upside-down V.

"Yoga came from the animals," Suzi told

me during a break.

"What do you mean?" I asked.

"Monks were in the forest looking at the ways animals stretch and move, and the monks started to copy them. Animals automatically know that they need to stretch and keep their hearts open. You don't really see a lot of animals with hunches in their back!"

"That's true," I said.

"I mean, sometimes you see squirrels and stuff. But animals instinctively know that they have to stretch. It's us humans that need the reminder."

Though some criticize doga for reducing an ancient tradition to a farce (one critic suggested that it must have been invented by "some really bored, unusually fit moron hipster"), Suzi and others defend it as valuable way to stay in shape — all the while bonding with your pets. "People include their dog in more and more activities, so why not in yoga?" she said.

That seemed perfectly reasonable to me, though Suzi conceded that it's easier (and probably more fun) to do doga with little dogs. I watched as she effortlessly incorporated Curli — who weighs eight pounds — into many of her poses, including lifting him high into the air for her Warrior Pose. Curli looked like he'd been there before, and he almost seemed to smile as the wind rustled through his shaggy white fur.

The next day, I took Casey dock jumping. The sport — which tests how far dogs can jump into a pool — is considered one of the safest for canines; there's little pressure on the animals' joints, and dogs land safely in water. There's also little doubt that dogs love competing. I've never seen as many wet, happy dogs as I did in the three hours I spent at the Ultimate Air Dogs event, which was held in a small Jacksonville park in the shadow of a highway overpass.

When Casey and I arrived, a handful of dogs were frolicking in a forty-foot pool. Music blasted from speakers. Other dogs barked and chased each other in the grass. I made the rookie mistake of telling Casey to "stop barking," prompting Ultimate Air Dogs founder Milt Wilcox — a big, hulking man in a red baseball cap — to utter three words I'd never heard before: "We encourage barking!"

It was Casey's lucky day.

A former major league pitcher who won a World Series with the Detroit Tigers in 1984, Milt said he fell in love with dock jumping thanks to his black Lab, Sparky. Any dog that likes water and has a "toy drive" can dock jump — even Bulldogs and Chihuahuas have competed at Ultimate Air Dogs events. But retrievers, Border Collies, and Belgian Malinois tend to dominate the sport. At the time of my visit, the reigning world champion was a five-year-old Chesapeake Bay Retriever from San Diego who'd jumped twenty-nine feet, one inch.

"Let's get Casey in the water!" Milt said, leading him up some stairs to the carpeted dock, where I'd watched Milt and others use the "chase" technique to get dogs to leap high into the air after a toy. (The dog would wait in a sit position while the handler walked to the end of the dock. The handler would then call the dog, who bolted toward him like a sprinter after a starting gun. When the dog was nearly out of running room, the handler would throw the toy high and out over the water, just in front of the dog's nose.)

"A dog's jump is only as good as his human's throw," said Victor Sparano, who was there with his wife, Susanne, and their year-old black Lab, Cooper. The dog hadn't been jumping that well on the day of my visit, which Victor and Susanne attributed less to Susanne's throws and more to the psychological impact of the relatively small crowd.

"Dogs are just like humans — most feed off the energy of the audience," Milt explained. "The more energy, the more screaming, the farther they jump."

Milt said that while a dog can improve his jump height and distance with training, a good jumper will instinctively leap into a pool after a toy. When I'd told Milt that Casey loves tennis balls and water, he'd practically guaranteed that my dog would be a dock jumping marvel. And though Milt tried valiantly to coax Casey into the pool, my dog — a relative old man at nine — was perhaps evidence that you can't teach old dogs new tricks. Casey eyed the two-foot drop between the dock and the water and then looked back at Milt, as if the discrepancy warranted some kind of explanation. Casey wasn't so sure about jumping off what probably felt like the end of the world.

He did eventually lower himself into the pool (carefully, front paws first) to retrieve the tennis ball. It wasn't a jump so much as a slow-motion flop, but it was a start. Casey got more daring as the day went on, especially when I got up on the deck with him. Dogs can be excellent copycats, so I threw a tennis ball in the pool, ran down the deck with great enthusiasm, and cannonballed into the water. A few minutes later, Casey did his best jump of the day. It was generously scored at four feet. The crowd went wild.

Maybe Florida wouldn't be so bad, after all.

After a few days in Jacksonville, I drove south to the Radisson Hotel in Melbourne Beach, host of the 19th Annual International Conference on Comparative Cognition. I'd never been to a canine cognition conference, and I was taken aback to learn that my dog wasn't welcome.

"I'm sorry, sir, but dogs can't be in the hotel," a Radisson employee announced as I strolled down a hotel hallway with Casey.

"But I'm here for the *animal cognition conference,*" I protested.

"I'm sorry," she said. "It's hotel policy."

"I hope you get the irony of a dog not being allowed at a conference partly about dogs," I said.

She stared at me blankly.

I walked Casey back to the Chalet, deposited him inside, and returned to the hotel with a chip on my shoulder. To take my mind off the indignity, I perused the conference program for panels I might want to attend. There were a dizzying number to choose from: "Attentive Spiders That Eat

Mosquitos," "Social Learning by Imitation in Bearded Dragons," "The Behavior of Asian Elephants During Mirror Exposure," and my favorite, "When Pigs Fly."

For some of the handful of canine cognition researchers in attendance, a more pressing concern than all that was the motivation of the American dog. Can dogs be said to *love* us? In *The Modern Dog*, Stanley Coren writes that "if you want to cause a commotion in a psychology department or any other place where animal and human behavior is studied, all that you have to do is to claim that your dog loves you."

But researchers across the world are struggling with variations of that question. Some, like neuroscientist Gregory Berns, flat-out equate dogs to people in this regard and insist that canines very much love us. Others don't go nearly that far.

"Do dogs hang out with us mostly because we're treat-dispensers, or can they be said to have a real and genuine affection for us?" wondered Erica Feuerbacher, a conference attendee and then a Ph.D. candidate at the University of Florida. We spoke on the hotel patio, only feet from the beach. "Has something changed along domestication from wolves where our interaction or our presence is rewarding to them in itself? We have these sayings in everyday life about dogs and unconditional love, and dogs being 'man's best friend,' but what produces and maintains that relationship? We don't really know."

To begin to explore that question, Erica devised an experiment to test whether dogs would work harder for a reward that consisted of food or human petting. The results were predictable — and depressing. Dogs want our treats more than they want our love.

"I used shelter dogs in my experiment, which we would think are deprived of human interaction and might really want love and petting," she told me at the conference. "But we found that dogs worked harder for food than human interaction." (The lone exception was a female Pit Bull, who responded equally to human interaction and to food.)

Erica then repeated the experiment with nonshelter dogs. "We wondered if maybe dogs had to have a history with a specific person for the human interaction to be meaningful," she said. "But we found the same pattern — dogs preferred food."

At a circular table in a Radisson conference room, several of Erica's fellow researchers from the University of Florida's

Canine Cognition and Behavior Lab waited to present their papers. Their canine studies were diverse, covering everything from social play to shelter behavior to odor detection. Nathaniel Hall, the author of a study about dogs and scent, told me he was interested in learning the best ways to train dogs for critical jobs, including detecting drugs and explosives. (A dog's nose is significantly more sensitive than a human one; while we have about five million scent-receptor cells, a German Shepherd has some 225 million.)

"There's still so much we don't know about a dog's sense of smell," Hall said, "especially when it comes to understanding how everyday odors influence our pet dogs' behavior. For example, pet owners assume that when a dog sniffs another dog owner for more than a few seconds, it's because the dog's interested in the smell of the other dog. We might say, 'Oh, your dog must smell my dog on me.' Whether that's true, I don't know."

Another Ph.D. student, Sasha Protopopova, was presenting a paper looking at whether shelter dogs trained to gaze into the eyes of humans were more likely to get adopted. Sasha's research focuses on canine adoption rates and adoptee preferences, and she wondered whether training dogs to make eye contact with humans would better their chances at leaving the shelter alive. In her study, though, Sasha found no statistical difference in adoption rates for shelter dogs trained to stare into our eyes.

Florida's Canine Cognition and Behavior Lab was led at the time of my visit by psychologist Clive Wynne, a former pigeon researcher considered by some to be the Debbie Downer of the canine cognition field. (He has since moved to Arizona State University.) Among those who study the intelligence and emotions of dogs, Wynne has perhaps been the most adamant about not looking through the lens of anthropomorphism.

"People may behave like animals," he once said, "but dogs are just good at being dogs."

Unlike Brian Hare, who runs the Duke Canine Cognition Center, Wynne would never write a book called *The Genius of Dogs* — he would argue that labeling dogs "geniuses" is just another way we project our human understanding of intelligence onto them.

"Our love for dogs can sometimes lead us — even those of us who are supposed to be guided by the science — to exaggerate just how much they truly grasp," Wynne told me during a break at the conference. "I'm certainly not trying to deny that dogs have an astonishing sensitivity to human beings, because they do. But I do try to be the voice of reason. I suppose that's earned me a reputation as a bit of a curmudgeon."

Though dogs' behavior can seem uncannily human, Wynne stresses that dogs have vastly different cognitive and perceptive abilities. "I understand why people sometimes think their dogs are incredibly intelligent, to the point of being able to sometimes read our minds," he said. "When you get up from your chair at home, sometimes you're headed to the bathroom, sometimes you're headed to the kitchen to make coffee, and sometimes you're headed to take the dog for a walk. Why does the dog so often stay lying down when you do the first two, but not the third? Is the dog a mind-reader? No. But your dog *is* a master at observing you and looking for any correlations between your movements and the crucial outcomes for the dog — being fed, going to the bathroom, that kind of thing."

Wynne believes that dogs learn by socialization and observation, and that there's not all that much difference between the brain of a dog and a wolf. Others, including Hare, fundamentally disagree, arguing that domestication has rewired the dog's brain and given them a remarkable ability to understand human gestures and cues.

"Brian and I have a massive disagreement," Wynne said, "and we could probably make some money by going on the road and debating this."

Wynne and Hare do agree on one thing, though: they're not all that interested in figuring out whether dogs are as smart as two-year-olds. When I visited Hare earlier in my journey at his Duke University lab, he told me that dog lovers often want answers to questions that can't really be measured.

"At my lab, we're not trying to understand if dogs are little people," Hare said. "I understand why people want to know if their dog is as smart as a two-year-old — the dog lover in me kind of wants to know that, too. But from an evolutionary perspective, that question doesn't make a lot of sense. If I had some magic way to transplant a chimpanzee brain into a dolphin and a dolphin brain into a chimpanzee, that wouldn't tell us much, because each brain has evolved to solve really different problems. The interesting question to me as a scientist isn't whether a dog is as smart as a kid. The interesting question is, Why are animals the way that they are, and how does evolution shape them to be that way? Basically, how does evolution shape cognition?"

One way in which a dog might be assumed to be more helpful than a two-year-old is in an emergency. Perhaps we have Lassie to blame for that — she could always be counted on to summon help if we found ourselves at the bottom of a well. But when William Roberts and Krista Macpherson from the University of Western Ontario tested how dogs reacted to two separate calamities befalling their owners, "not a single dog did anything useful at all," Wynne told me.

The study, titled "Do Dogs (*Canis familiaris*) Seek Help in an Emergency?," is a cautionary tail of high canine expectations. In the first of two experiments, twelve dog owners walked their dog on a leash to the middle of a field and then pretended to keel over from a heart attack. A stranger sat ten meters away, and a video camera in a tree recorded the proceedings. The results? A few of the dogs pawed their owners before appearing to lose interest.

One dog, a toy Poodle, did approach the stranger, but "it ran over and jumped in the person's lap — not because it was trying to signal that its owner was in distress, but because it wanted to be petted," Alex Boese suggests in his book *Elephants on Acid: And Other Bizarre Experiments.* "It probably figured, *Uh-oh! My owner's dead. I need someone to adopt me!*"

In a related experiment, Roberts and Macpherson had people walk their dogs into a room (where they were greeted by a stranger) and then proceed to a second room, where they'd been instructed to pull a bookshelf down on top of themselves in a way that looked like it really hurt. The humans then screamed in pain and implored their dogs to get help. Once again, man's best friend didn't appear especially bothered.

"In no case did a dog solicit help from a bystander," the researchers found.

There are several possible conclusions to be drawn: One, dogs don't care if we die. Two, dogs don't understand the meaning of "getting help." Three, dogs are actually too smart to be fooled by fake heart attacks and slow-falling bookshelves. (If dogs can distinguish between when we're getting off the couch to go to the bathroom and when we're getting off it to take them for a walk, surely they can distinguish between a real accident and playacting.)

Whatever the case, Brian Hare and

Vanessa Woods caution us in *The Genius of Dogs* that though dogs are "arguably the most successful mammal on the planet, besides us," they have many blind spots. Dogs are terrible at physics, for example. They don't understand basic spatial and connectivity principles, including "that a ball cannot pass through a solid object like a wall, and that when two toys are connected, if you move one, the other will move, too."

The Genius of Dogs includes a handy chart comparing the cognitive abilities of canines to other mammals. The authors give dogs "genius" rankings for "comprehending visual gestures" and "learning new words," but they're only average at "navigating through space" and "vapid" when it comes to "understanding physics."

Though dogs will usually take the shortest route to a reward, things get dicey when they face detours or barriers. Unlike wolves, dogs aren't good at adjusting mid-route. They get easily stumped in a maze, developing an unsuccessful strategy "where no matter what they could see in front of them, they alternated between turning left and right." (Though everyone seems to know a story of a dog who gets lost and then finds his way home, canine cognition researchers insist that it's rare — and that you should probably just have your dog microchipped instead.)

Worst of all, dogs seem to have little awareness of their own cognitive limitations. If you hide food from a chimpanzee, Hare and Woods write that it "will inspect different hiding locations before making a choice." Dogs, on the other hand, just guess.

"Studies designed specifically to examine whether dogs know if they know something have found little evidence that dogs are aware of their own ignorance," the authors conclude.

I'm not so sure what to make of that. After all, the same can be safely said of some humans I know.

One of the challenges of driving around the country for months in an RV is that seemingly everyone — friends, Facebook followers, long-lost lovers — expects a visit.

"Oh, I'm *barely* out of your way," they would lie, bribing me with promises of home-cooked meals, charming guest cottages, or reunion sex.

To many people, a motorhome conveys the kind of vehicular lollygagging that lends itself to day- or week-long detours. But I had a limited amount of time in the Chalet, and I decided that any significant deviation from my itinerary had better come with the promise of a good dog.

I got one such offer in Florida, and it came at an opportune time. After a week without Sam in the motorhome, I was feeling lonely. So I was happy to drive to Fort Lauderdale to visit my newly engaged gay twentysomething friends Neil and Brant and their dog, Amelia. She'd been found wandering through a Winn-Dixie parking lot five months prior to my visit. She ended up at the Humane Society of Broward County, where shelter workers temporarily named her Dixie. As best as anyone could tell, she was a Rhodesian Ridgeback/Lab mix. (She probably also had some Pit Bull in her, but shelters across the country go to extraordinary lengths to pretend Pit Bulls aren't Pit Bulls.)

Neil and Brant had showed up by chance at the shelter the day after Dixie arrived. The couple hoped to adopt a child together eventually, but first they wanted to test their parenting skills on a nonhuman. They figured they would adopt a small dog, because they lived in a modest apartment complex while Brant attended medical school.

But then they met Dixie. When they first

spotted her at the shelter, she was napping on a cot in her kennel. Though Neil and Brant remember different details from that moment, they agree that she eventually noticed them staring at her, jumped up from her cot, and walked over to say hello through the kennel fence.

"We melted," Neil recalled. "It was pretty much decided right there that she'd be coming home with us."

After waiting three days (Dixie was placed on a temporary hold in case anyone claimed her), Neil and Brant took her home and renamed her. Amelia quickly ingratiated herself with the couple — she watched television with them on the couch and hogged their bed at night. They took her for long walks and let her race after other dogs at the local dog park.

But Amelia came with some quirks. She would cower in fear, for example, whenever Brant or Neil picked up a shoe. If they left their shoes out, she would destroy them. Though crate-trained, Amelia once broke out when Neil and Brant weren't home and ransacked their apartment. Brant showed me a picture of the damage. I'd never seen anything like it — the room looked as if an angry mountain lion had recently come for a visit. There was nothing left of the comforter. The floor was strewn with shirts, boxer shorts, a broken cologne bottle, and white pillow stuffing. A black dresser leaned on its side, defeated. The window shade was slashed.

"She has just a *tiny bit* of separation anxiety," Brant joked as Amelia tried to get Casey to chase her around their apartment on my first afternoon there. "The few times we've let her have the run of the apartment, we always come back to find something destroyed."

Before going to dinner that night at a surprisingly decent sushi buffet, I suggested we put a ThunderShirt on Amelia. "It didn't work for Casey," I said, "but I think he might be an outlier." Neil and Brant discussed it and decided to be brave. "Here goes nothing," Neil said, as I helped him wrap Amelia. "We'll let her have the run of the apartment and see what happens."

When we returned ninety minutes later, Amelia was hanging out on the couch, seemingly without a care in the world. We inspected the apartment for damage. Nothing.

Brant screamed, "It's a miracle!"

The next day, we took the dogs to a local dog park and then came home and relaxed by the pool. After nearly ten days in the sun,

I was looking positively Mediterranean. Neil and Brant caught me up on their wedding plans and told me about the quaint, dog-friendly bed and breakfast in Vermont where everyone would be staying.

Though I was genuinely excited for them, all their happy talk and tender physicality made me envious and more than a little sad. I'd hoped to marry my ex-boyfriend, but that had become unlikely since our breakup.

To take my mind off my feelings, I spent that evening flirting with a friend of Brant's from medical school. Marc was twenty-eight, handsome, charming, and smart. Best of all, he said yes when I offered to give him a tour of the Chalet.

"Guys have invited me back to their apartments and their cars," he said, "but never an RV!"

We talked for an hour that night in my motorhome — about Florida, medical school, the trip, the book, and the Facebook page. As the evening turned to morning, I didn't want him to leave. Neither did he.

"Of course this happens," he said before heading home. "I meet someone great, and he doesn't live in Florida and is leaving tomorrow to drive around the country."

Before I could stop myself, I heard myself say, "Maybe you could join me for some part of the trip?"

Little did I know what I was getting myself into.

Casey and I spent our last two days on Florida's Gulf Coast.

First we stopped in Sarasota to meet Cary, a woman who'd read about my journey and suggested that I come meet her black Lab, Pepe. Cary and her husband had adopted the dog a few years before, and though he'd arrived housebroken and well socialized, he wouldn't respond to basic commands like "Sit" or "Stay."

Cary suspected that the dog had been a member of another family and most likely had a name, but she'd had no idea what it was. "I would sit around with the dog and toss out names like Rover and Blackie, just to see if we might get lucky," Cary told me, relaxing in a squeaky rocking chair in her living room overlooking her condo association's man-made lake. Her husband, Mike, swayed in his own rocking chair across the room.

None of the dog names worked until the day Cary, who is Cuban, tried traditional Hispanic ones. "One day I'm sitting in the living room with Mike, and the dog is lounging on the couch," Cary recalled. "On a whim I said, 'Pepe,' and the dog jumped off the couch and came over to where I was sitting. My husband and I looked at each other and said, 'No way!' "

They tried again a few minutes later with the same result. Cary then told Pepe to "Siéntate," a command that the dog had always ignored in English. Pepe sat right down. "I started going through the basic commands in Spanish, and Pepe knew all of them," Cary told me. "I wish I could have taken a picture of his face. He was ecstatic. His humans were finally speaking his language!"

Cary had already bonded with Pepe before learning his name, but she and Mike credit speaking Spanish with deepening their connection to their dog. Though he now knows commands in English as well, Cary says Pepe is still more likely to respond to Spanish.

"And he still barely listens to me," Mike lamented. "He looks at me while I butcher Spanish, and I swear he's thinking to himself, 'What the hell is this imbecile saying?' He definitely likes Cary more than he likes me."

I told them about Piper, a dog I was going to meet that afternoon in Tampa. Piper had bitten a home intruder two years prior, only to have the robber stab her with a crowbar.

"Poor dog," Cary said.

I brightened her mood by filling in the rest of the story. "The day after the robbery," I explained, "a police dog named Bosco chased down the suspect. Bosco even roughed him up, for good measure."

Cary's face lit up. "Serves him right to mess with dogs!" She looked lovingly at Pepe, who loafed on the couch, his pink tongue dangling from his mouth. "If someone stabbed my dog, that would be like someone stabbing my child. To me, my pets are like my children. I love them the same."

Although I adore dogs, I'm surprised when I hear people equate their love for their pets with their love for their kids. Did Cary actually mean to say there's *no difference* between the depth of love she feels for her daughters and the affection she has for her dog? If so, was that a sign of some advanced and egalitarian perspective on the value of different species? Or was it a sign of insanity? Whatever the case, did Cary really want her daughters to see that in print?

"I don't have kids," I told her, "but I imagine that I would love them in a different way than the dogs I've loved in my life."

275

I suggested she think about how she might mourn the death of a child. "If Pepe died, do you think you would feel just as sad as if one of your daughters passed away?"

Cary considered the question for a moment. Then she told me about Alf, the dog she'd owned when her girls were young. "Raising Alf was like raising a child," she said. "He was very much a kid, just like my daughters. He was part of the family unit. We had a big Fisher-Price slide in the living room back then, and he would go down the slide, just like my girls did. I'd give them popsicles, and he would stand in line with my girls and wait for his. So when I had to put Alfy to sleep . . ."

Cary started to tear up. "I still get emotional, as you can see. It was probably the worst thing I've ever had to do."

Though Cary couldn't tease apart her love for her dogs and her children, she was quick to differentiate between how kids and dogs love their human caretakers. "Dogs never become teenagers," she told me. "It's a consistent relationship; the quality of their love for you doesn't change. They don't grow up and tell you that you're the worst. They don't move out. Even if you screw up, they don't hold it against you."

Mike chimed in. "Dogs are like kids, but

without the drama or the attitude."

(I'd heard something similar at a KOA in Florida. An elderly couple had invited me inside their mammoth motorhome and explained why they liked spending most of the year on the road with their two dogs. "Frankly, we do this so our kids won't know where to find us," they'd said, grinning like schoolchildren.)

Cary said she'd learned a lot about human family dynamics in her years as a midwife and a nurse. "I've seen a lot of bitterness and downright hatred between family members who were supposed to love each other," she told me. "But with dogs, you don't seem to have that. They're always going to be there, always going to want you."

Cary went on. "A dog's love is different from human love, because it's truly unconditional."

Is it, though? Patricia McConnell writes that "it's become a cliché that we love dogs because they give us unconditional positive regard." McConnell calls this belief "naïve" and suggests we've "convinced ourselves that our dogs love us constantly and relentlessly, simply because we're not very good at reading their nonverbal communications to us." McConnell writes about her Border Collie, Cool Hand Luke, who once saved her life but can still shoot her "a look that can only be translated as a four-letter word."

Anthony Podberscek, who studies the human-animal bond, told writer Hal Herzog that the unconditional love theory of pet ownership is "rubbish," a peculiarly American phenomenon that's dismissive of the intelligence and emotional complexity of pet animals. Podberscek argues that if we truly believe "that our pets are programmed to mindlessly love us no matter what we do to them," Herzog reports in his book *Some We Love, Some We Hate, Some We Eat,* "they are essentially Cartesian robots that take whatever we dish out for them and then come back for more."

Still, I could understand where Cary was coming from. Adolescent dogs are less likely to *"hate you"* than adolescent humans. As psychiatrist and educator Aaron Katcher once wrote, "a dog is like an eternal Peter Pan, a child who never grows old and who therefore is always available to love and be loved." Andy Rooney may have pinpointed another advantage that dogs have over kids: we don't have to listen to dogs complain.

"If dogs could talk," Rooney once said, "it would take a lot of the fun out of owning one."

■ ■ ■ ■

I stayed on the topic of kids and dogs later that afternoon in Tampa, when I visited a stay-at-home mom named Kim and her dogs, Piper (the mutt who'd been stabbed by the intruder) and Hunter (a Brittany). We sat with Kim's mother and daughter on the back patio overlooking their pool and the ball fields of an adjacent middle school. As we spoke, two squirrels tortured Piper, Hunter, and Casey by racing back and forth on top of a wooden fence. Next door, a neighbor's parakeet did a perfect impersonation of a ringing telephone.

After I told them about my conversation earlier that day in Sarasota, Kim spoke about the pain of losing her previous dog, Scout, to leukemia. "I had to put him down, and it felt to me very much like losing a child," Kim said.

"Were you especially close to Scout, as compared to other dogs you've lived with?" I asked.

"Yes, and I do think it's true that we have deeper connections with certain pets than with others," she told me. "That dog used to go everywhere with us — we'd take him on trips, he'd ride with us on our motorcycle. If Scout couldn't come somewhere with us, we didn't go."

A few minutes later, Kim's mom, Carol — a youthful and funny psychotherapist in her sixties — recalled the pain of losing her favorite dog, Shoshanna. "She died a few years after my husband passed," Carol said, "and what made it especially painful was that she was my connection to my husband. It was his dog, his love. I mean, he let that dog sit *on* the table at dinner! When Shoshanna died, it brought up all the feelings of losing my husband again."

But Carol was sure to distinguish between the pain of losing a dog and the agony of losing her daughter. "My daughter died when she was thirty-seven, and *nothing* compares to that pain. If you can picture someone taking your guts, pulling them up through your throat and out your mouth while you're still alive, that doesn't even compare to the pain you feel. If someone tells you that they feel the same losing a child as they do losing a dog, then they haven't experienced the death of a child."

At that moment, Casey, who had been wandering through the yard, spotted an armadillo on the other side of the chain link fence that separated Kim's property from the school's fields. He let out two barks in quick succession and rushed toward the fence, but he could only watch as the armadillo — moving with surprising deftness for an animal carrying an armored shell — scurried away.

"It's certainly the natural cycle for dogs to die before us," Carol continued, "and in many ways they teach us how to deal with loss. But we're not supposed to outlive our children. That's not the natural order of things. I love my dogs, and they are my best friends. But I don't dream about them when they die. I mourn them, and then I move on and get another dog — another best friend for the next ten or fifteen years of my life."

I looked at Casey, who had forgotten about the armadillo and was now busy sniffing a bush. It occurred to me — I mean, *really* occurred to me — that he wouldn't live forever. I felt a tightness in my chest, a kind of panic.

"I don't want Casey to die," I heard myself saying out loud, though I'd only intended the words for myself.

Then I thought about what might happen if I died before Casey. Who would look after him? I hadn't prepared a will, hadn't made any arrangements. But those were mere practicalities. Where my mind went, I'm ashamed to admit, is toward more ego-based questions: Would Casey even notice that I was gone? Would he be sad? Would he mourn *me*? The answers seemed painfully clear to me that afternoon in Florida: maybe, probably not, and no.

Oh, how part of me yearned for a dog like Capitán, the German Shepherd who spends most of his days sitting by his master's grave in Argentina. Or Ciccio, another German Shepherd who waits for his owner each day outside a village church in southern Italy. For years, Ciccio's owner, Maria, would walk her dog to a church in San Donaci every day as the tower bells began to chime. When Maria died in 2012, the dog first participated in her funeral procession, then continued to make his way to the church each day as the bells started chiming. The village has since adopted Ciccio, collectively feeding and sheltering him, but as he sits outside the church each afternoon, it seems he's still devoted only to Maria.

Perhaps I would have been better suited for the Victorian era. As Susan Orlean recounts in *Rin Tin Tin,* dogs of that period were believed to be "indefatigable mourners. They were said to visit their masters' graves on their own, lying on the freshly turned dirt for days, inconsolable. If their grief was too much to bear, dogs sometimes committed suicide; newspapers of the period carried frequent reports of these canine deaths. One of the great attractions of having a pet, then, was believing it would miss us and mourn us and always remember us, even if friends and family let us down."

But as Orlean points out, this Victorian obsession crowded out the reality of sharing one's life with a dog — unlike with human children, we usually outlive our *furbabies*. "A dog's life is a short one," she writes, "so most of the time it is we who are mourning them."

The French call dogs "bêtes de chagrin" ("beasts of sorrow"), because, as writer Roger Grenier puts it, they "inflict the suffering of loss upon us."

I hoped to have a few years before Casey inflicted the suffering of loss upon me. More than anything, I hoped I wouldn't regret the life I'd offered him.

Casey, in the RV on Day 1 of our voyage.
Everything was fine until I started the engine.

An English
Springer
Spaniel at the
Westminster
Kennel Club
Dog Show,
considered the
Super Bowl of
dog breeding.
(Brad DeCecco)

Casey and I during a break from the human drama at Tompkins Square Dog Run in
Manhattan. *(Brad DeCecco)*

Kim Zakrzewski with Baxter, a Westie/
Bichon Frise mix. Kim was taken to
court for allegedly not picking up after
the dog.

"Wolf Man" Rob Gudger and his
wolfdogs in their pen behind his house
in the North Carolina mountains.

It was all fun and games at this Ultimate Air Dogs dock-jumping event in Jacksonville,
Florida.

Hanging out with Casey—
and looking oddly like
Edward Norton—at an RV
park in St. Augustine, Florida.

I spotted many funny signs on my journey, including this one at the Pecan Grove RV
Park in Austin, Texas.

This mangy mutt got a second chance thanks to the rescue organization San Antonio Pets Alive!

Casey at the Marfa Lights Viewing Area in Marfa, Texas.

Cows were no match for two Border Collies on a cattle ranch in Gunnison, Colorado.

One of the stray "rez dogs" I came upon on an Indian reservation in Arizona. I named her Rezzy and decided to take her along for the ride.

The RV parked outside my father's house in the Arizona desert.

Casey and Rezzy
across the bay from
San Francisco, where
I was born and
raised.

"Dog Whisperer" Cesar Millan gave
Casey, Rezzy, and me a tour of his
Dog Psychology Center—and told
me everything I was doing wrong
with my dogs.

In Kent, Washington, I spent time
with this homeless teenager and her
then boyfriend—and their two dogs.

Pet photographer Amanda Jones photographed the dogs and me in San Francisco. Notice me holding Rezzy—unlike Casey, she wasn't a "model model."

(Amanda Jones)

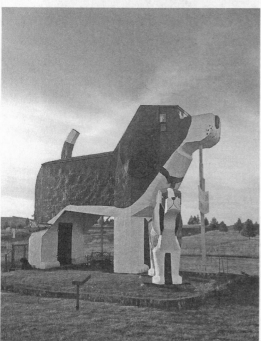

The bathroom is in the rear at the Dog Bark Park Inn Bed & Breakfast in Cottonwood, Idaho.

Dog trainer
David Riggs and
his Lab, Rocky,
in Montana.

Casey and I hiking in
Gunnison, Colorado.

Dog rescuer Randy Grim, founder of Stray
Rescue of St. Louis, rescued an emaciated
Pit Bull in East St. Louis.

PART TWO

6.
IN WHICH I CRY OVER DOGS IN WEST TEXAS, HIRE A BED BUG EXTERMINATOR IN NEW MEXICO, AND MEET GAY COWBOYS IN COLORADO

It's clichéd, of course, to blast Willie Nelson's "On the Road Again" while rolling along a lonely two-lane highway under a sun-drenched Texas sky, with only circling vultures and a desolate landscape for company. But that doesn't make it any less satisfying.

It was late March, six weeks since leaving Provincetown, and Casey and I were headed west on Route 90, which hugs the Mexican border west of San Antonio. We were headed to Marfa, a small town in the high desert named after a character in Dostoyevsky's *The Brothers Karamazov.* On the brink of becoming a ghost town in the early 1970s, Marfa had begun its transformation with the arrival of minimalist artist Donald Judd, who bought up downtown real estate and turned the 340-acre Fort D. A. Russell military installation into a contemporary art museum. Today, Marfa is home to an eclectic collection of artists, retired ranchers, and Mexican Americans — including the town's elderly dogcatcher, who won the lottery. The area also boasts the Marfa Lights, mysterious brightly colored spheres that can sometimes be seen at night on the southwest horizon.

I'd never been to South Texas, and I was struck by the area's barren beauty. The landscape along Route 90 came with surprises, like when we passed over the majestic Pecos River, which empties into the nearby Rio Grande. I stopped to have lunch — and let Casey pee — at a scenic overlook with a stunning view of the river's high canyon walls.

After rolling through Langtry, the long-time home of famed judge/saloonkeeper Roy Bean, there was nothing approximating civilization for the next 150 miles. We came upon an occasional big rig or Border Patrol vehicle, but for the most part we had Route 90 to ourselves. The road to Marfa was straight, flat, and seemingly endless.

Casey spent most of the drive curled up and snoozing in his usual spot between the driver and passenger seats. To my delight, he'd grown increasingly comfortable in the RV during our quick drive through southern Alabama and Mississippi. Sadly, I couldn't say the same for myself. My butt had started hurting by the time I'd reached Louisiana, where I visited Animal Rescue New Orleans, a shelter and adoption organization. Every few minutes, I'd adjust my sitting position to combat the soreness that comes from six weeks behind the wheel.

I'd had some human company in New Orleans. Marc, the guy I'd met less than a week before in Fort Lauderdale, had flown out to see me. To our credit, we both recognized that we were probably out of our minds. After only a brief encounter in a motorhome, we found ourselves walking together down Bourbon Street, feeding Casey beignets, and generally reveling in that delirious first stage of budding romantic relationships — infatuation.

But we took comfort in the fact that we could speak soberly about our recklessness. We weren't just two lovestruck idiots; we were two *self-aware* lovestruck idiots. We spent a lot of time in the Big Easy — and on the phone before and after our time there — psychoanalyzing our intense mutual attraction.

When we weren't navel gazing, we spoke about our families. At the time, Marc was mostly estranged from his. He'd never been close to his father. He had been close to his mother, but when he told her, in college, that he was gay, she insisted that he was accepting the devil and announced that he was no different from the criminals who hung next to Christ on the cross. Marc stopped going home for holidays after that.

I told Marc about my father, who had come a long way since learning I was gay during a summer break in college. "I guess this is what I get for raising you in San Francisco," he'd said, teary-eyed, when I broke the news. Though my dad isn't a religious man, he'd struggled at first with the usual suspects: Was I destined for a difficult life? Would I contract HIV? (When I came out to him in 1995, an HIV diagnosis was still considered a death sentence in America.)

It was fascinating — and heartening — to watch my father's transformation over the next decade. At first, he struggled to say the word "boyfriend." Whoever I happened to be dating was my "friend" — as in, "How's your friend doing?" "What's your friend up to these days?" Gradually, his perspective and vocabulary changed. Today, he's the model father of a gay son and encourages me to bring whomever I'm dating home for the holidays. I wondered if Marc might eventually be a candidate.

Privately, though, I worried about whether a whirlwind romance might distract me from the task at hand: traveling around the country with my dog. I'd designed the journey to strip away all non-dog-related distractions, and now here I was with a schoolboy crush. I'd also promised myself that I wouldn't run away from feelings of loneliness that I knew would surface during four months on the road.

I had planned to "lean into" any uncomfortable emotions, as Buddhist nun Pema Chodron suggests in her books, including *Comfortable with Uncertainty* and *The Wis-*

dom of No Escape. I'd brought along a handful of her texts, just in case I needed the reminder. Loneliness, she assures her readers, is not the enemy. Like any unpleasant feeling, it can instead be "the perfect teacher." That is, if we let ourselves feel it instead of doing what we usually do — diving headfirst into distraction.

But dive I did. I called Marc every day from the road. In between calls, I'd text him pictures of Casey and other dogs I'd met; he'd send me back adorable photographs of himself studying or getting out of the shower. There's nothing like pictures of a person you're newly enamored with to get you through long stretches of Louisiana and Texas.

After saying good-bye to Marc in New Orleans (and agreeing that we would see each other again a few weeks later in San Francisco), I'd spent a week putzing around Texas. I loved my three days in Austin, a beautiful and quirky city that in 2000 adopted the slogan, "Keep Austin Weird." Casey swam near Barton Springs, and we settled in for a few days at the city's Pecan Grove RV Park, where a black and white Pit Bull hung out under a sign that read, "Life is too short to live in Dallas."

■ ■ ■ ■

Just when I began to wonder if Casey and I might be the only living things on Route 90 on the way to Marfa, we came upon a temporary interior Border Control checkpoint manned by three uniformed officers and a beautiful German Shepherd. They stood in the shade under a small canopy.

"Where you headed?" asked the youngest of the three, who looked like a truant, as if he'd skipped high school that morning to play border agent with his buddies. Perhaps to compensate for his baby face, he puffed out his chest and tried his best to seem imposing. As he spoke, the German Shepherd gracefully lifted himself on his hind legs and rested his big front paws on the still of my open window. Before I knew what was happening, the dog's wet nose was practically in my lap.

"On my way to Marfa," I told the officer, trying not to sound like the kind of person who might be smuggling drugs, explosives, or a family of illegal immigrants.

I had an urge to pet the animal but thought better of it. "I suppose I'm not supposed to touch the dog?" I asked.

"Correct," the officer said.

Had Casey been awake, he likely would not have reacted well to this strange dog's head protruding through our window. Luckily, he slept through the whole thing. When the German Shepherd was done sniffing, the dog gently pushed back from the sill and returned to ground level.

"You're all set," the officer said, waving me along with an uninterested flick of his wrist.

After driving through the juniper shrubs and desert canyonland of Paisano Pass, a gap at five thousand feet above sea level, Casey and I rolled into Marfa under a passing rain shower. We'd encountered little rainfall to that point on our journey. The worst storm had come a few days before in Houston, where I parked the Chalet in an RV park next to the husband and wife team of Amy and Rod Burkert, who live full-time in a Winnebago with their two dogs and run a website — gopetfriendly.com — for people who travel with their pets. That day, a severe weather advisory urged people to take shelter inside "sturdy structures." I didn't know if a motorhome counted as "sturdy," but I held Casey close as heavy winds shook the RV and lightning bolts lit up the night sky. I even mouthed a panicked prayer: *Please, God, don't let us die in a mobile home in Houston.*

I pulled off Route 90 into Marfa's Tumble In RV Park, which was deserted except for an orange and white 1962 Mobile Scout trailer that served as its unmanned office. A sign explained that guests were to leave cash, a check, or credit card information in an envelope and drop it in the payment box. Call it the Far West Texas honor system.

I took Casey for a walk toward nearby train tracks, but we couldn't get far without stepping on goathead thorns, or what some locals call Texas sandburs. I pulled two of the small, prickly white thorns from Casey's paws (and dozens from the bottom of my shoes) before cursing the plant, cutting our walk short, and returning to the RV.

The sun was about to set, so I drove us a few miles east of town to the Marfa Lights Viewing Area, which is centered around a clay-colored circular structure and a large patio facing southwest, toward the Chinati Mountains. I parked the Chalet on the side of the road, some thirty yards behind the only other vehicle there, an absurdly long motorhome occupied by a couple from New Jersey (and their cat).

I introduced myself by asking for a favor. "Might you have a bottle opener I can borrow?" I said, standing on the pavement by their driver-side window with a chilled Corona in my right hand. I'd somehow managed to lose two bottle openers during my journey, but I was hell-bent on drinking a beer while watching this sunset — and, with any luck, spotting the Marfa Lights — from the roof of the RV. (During his time on the road with me, Sam had made a habit of climbing up the motorhome's back ladder and drinking up there with his shirt off.)

"I'm not sure I can, in good conscience, lend my bottle opener to a Corona drinker," said the jokester from New Jersey, as his cat stared blankly at me from her sprawled-out position on the dashboard. "Corona's not real beer."

"Don't listen to him," his wife interjected, playfully swatting her husband on the shoulder from the passenger seat. She disappeared for a few seconds before returning with an opener.

"Have you guys ever seen the lights?" I asked.

"Never," the man said. "We've been to this part of Texas a lot — we love how peaceful it is. We've spent a few nights here over the years and never had any luck."

He leaned out the window and pointed toward the southwest horizon, in the direction of Mitchell Flat and Route 67. "They'll be over there, if we see them tonight," he went on, cracking open a dark beer of his own. "You'll also see car lights, and sometimes people think those are the Marfa Lights — but those aren't them."

"When do they usually appear?"

"I've heard some people say that if you don't see them an hour or two after dark, then you won't," he said. "Other people say you should stick around until eleven. But I'm usually passed out drunk by then."

The lights have been a source of intrigue since at least the late 1800s, when settlers wondered if they might be caused by Apache campfires. In the years since, as the lights have drawn tourists from all over the world, dozens of theories have been proposed to explain the mystery — "from scientific to science fiction," according to a plaque in the viewing area. "Some believe the lights are an electrostatic discharge, swamp gases, moonlight shining on veins of mica, or ghosts of conquistadors searching for gold," the plaque goes on to explain.

Armed with two Coronas and a hastily arranged cheese plate, I climbed up the Chalet's back ladder to do my own Marfa Lights investigating. First, though, I was struck by less mysterious natural wonders — a rainbow to the southeast, and, minutes later, what writer Michael Hall once described as "one of those alarming West Texas sunsets, the kind that looks like somewhere the world is coming to a fiery end." Through distant electrical lines, the horizon looked bright orange, red, and, further up in the night sky, dark purple.

I spent the next two hours on top of the RV, staring out into the quiet, starry night. It was chilly now that the sun had set, so I pulled my sweatshirt hood over my head and stuffed my hands in my jean pockets. Every once in a while I would see what I presumed to be car headlights in the distance on Route 67, but I didn't see any lights dancing, breaking apart, or melting together, as those who have spotted the mysterious sight report.

What I did see while sitting in silence on the roof of my motorhome, what kept flashing in my mind's eye, were images from my visit two days earlier to San Antonio's largest shelter, Animal Care Services. Though I'd somehow managed the drive to Marfa without tearing up at the memory of my time in the shelter's euthanasia room, the honeymoon was over.

Back in 2004, Animal Care Services had been named the country's deadliest. "Simply entering virtually assures an animal's doom," the *San Antonio Express-News* wrote at the time.

The shelter had made remarkable progress since then, working with outside rescue groups — including San Antonio Pets Alive! or SAPA! — to increase adoptions. The shelter hoped to get its euthanasia rate (30 percent at the time of my visit) down even further and join Austin as a no-kill community, one that euthanizes only vicious or severely ill animals.

If it weren't for the outdoor livestock area, Animal Care Services could be mistaken for a large, modern elementary school. There are ten one-story buildings spread over several acres, and the shelter has plenty of grass and open space. My tour guide for the afternoon was shelter spokeswoman Lisa Norwood, who pointed toward a big brown horse standing in an outdoor corral.

"Welcome to Texas," she said, "where we get calls about stray horses literally walking down a street." Animal Care Services routinely rescues animals you might not expect: two weeks before my visit, it had taken in a stray potbelly pig.

A colony of feral cats also lived on the property. "If you see kitties, don't worry — they're not escapees," Lisa said with a smile.

"They're kind of all over the place, so I always have to point it out to visitors because they think the cats have just escaped."

Lisa saved the hardest part of the tour for last. When we entered the open-air euthanasia room (known as the EBI, for euthanasia by injection), a male employee was injecting a sedated white Pit Bull with sodium pentobarbital solution on one of three stainless steel injection tables. At the end of the row of tables, a conveyor belt led to the back of a dump truck. (Normally the shelter cremates dogs in an on-site crematory, but when the crematory needs maintenance, euthanized dogs are transported to a landfill instead.) Next to the conveyor belt, six dogs waited in crates on the epoxy-coated floor.

Ellen Jefferson, the founder of SAPA! and Austin Pets Alive!, stood with her back to the injection tables cradling a tiny black puppy in her arms. The six-week-old dog had distemper and was having trouble breathing, and though one of his siblings seemed to be recovering from the illness in foster care, this dog wasn't going to make it.

"We're going to have to put her down," Ellen told me, unusual words from a woman who has devoted most of her adult life to keeping dogs and cats alive.

I looked back at the Pit Bull, now dead. It had been friendly and healthy, I was told, but it had the misfortune of being a big dog — and a Pit Bull, no less — on a day when neither Animal Care Services nor SAPA! had space for her.

"I don't have room for any big dogs today," SAPA!s Holly Livermore told me earlier in the group's small shelter office, where fifteen dogs she'd already saved from that day's euthanasia list waited to be transported to foster homes or off-site adoption sites. All the dogs weighed less than thirty-five pounds, including an adorable black and brown Shepherd mix I'd taken to the shelter play area.

"You sure you don't have room for one more in your RV?" Lisa said, noticing how much I liked the dog.

Small dogs have two advantages when it comes to staying alive in America's shelters: they take up less space, and they're more likely to get adopted quickly. "There are so many great big dogs that we can't save in San Antonio," Holly told me. "I basically have to look at them and then look away, because if you look into their eyes too long, it's too much to handle."

Like many other Southern communities,

San Antonio has a pet overpopulation problem. Lisa estimated that some 150,000 dogs roam the city at any one time. Despite their effectiveness in other areas, spay and neuter campaigns are sometimes met with resistance in some parts of the South, where, as writer Hal Herzog put it, people "don't like restrictions on their dogs any more than they like zoning, gun control, or laws that keep you from carting little children around in the back of an open pickup truck."

Most adoptable dogs in Northern shelters, particularly in the Northeast, have been saved from Southern states and trucked or flown in by a vast and dedicated army of rescue groups. But those groups can't save every adoptable animal, including the six crated dogs in the Animal Care Services EBI. They were big animals — Labs, Pit Bulls, mixed breeds. Most had been found roaming the streets of San Antonio.

"They're not what you'd consider hardened street dogs," Lisa explained. "They're just dogs who happen to roam the streets because their owners let them. When the dogs get picked up, no one comes to claim them. It's sad, because most are sweet as can be."

I knelt down in front of their crates and

made the mistake of looking into their eyes. Two wagged their tails at me. One, a skinny young Pit Bull mix with a cut on her face, lifted her left paw, as if to say "Hello? Goodbye? Please don't let them kill me." Two of the dogs were owner-surrenders, meaning their caretakers had dropped them off at the shelter. I'd watched a dozen people relinquish their dogs to Animal Care Services. They all had their reasons — lost jobs, impending moves, husbands who didn't want the dog around anymore. Some cried as they handed over their pets. Others didn't. One said she'd happened to find the dog on the street, that it was a stray. But the shelter staff suspected she was lying.

SAPA! and other rescue groups do their best to convince folks not to leave their dogs at the shelter, offering everything from low-cost health care options to dog behavior resources. Many people who surrender their dogs don't know the basics of what a dog needs to be happy and healthy; they're ignorant about canine "health, behavior, and nutrition," Michael Schaffer writes in *One Nation Under Dog.* In a summary of owner-surrender studies, Schaffer found that "people giving up their pets displayed a degree of ignorance about animal brains, like believing they misbehaved out of spite."

When a teary, middle-aged Hispanic woman had brought in a mange-ridden black mutt, I watched as Holly promised that SAPA! would help pay for the dog's medical expenses if the woman would reconsider. She wouldn't.

"If we were in Austin, I would take this dog in a heartbeat," Holly told me. "We have an awesome medical facility in Austin, an awesome foster base. Mangys are my favorite kinds because you can turn them around, you can get them healthy and feeling better. But I can't save this dog today. I just can't."

Holly turned toward the dog, who was lying on the porcelain tile floor, looking both pathetic and, somehow, proud. It had, after all, survived a family that had practically let it die. "I just can't," Holly said again, seemingly trying to convince herself that she would need to give up on this one.

In the EBI, the next dog to die was a black Lab. A stray, he looked a bit like Blind of Arden, a slender and athletic retrieving marvel who appeared on the cover of *Life* back in 1938. Two shelter employees lifted the sedated animal onto the table, rolled him onto his side, and injected him with a dose of lethal serum. One of the men, reacting to an unrelated joke his colleague had uttered, laughed as he stuck the needle into the dog. The joke, the laughter — maybe they were necessary coping tools for those who kill healthy dogs for a living.

I found it impossibly cruel that the crated dogs had perfect views of the conveyer belt. Every few minutes, a canine corpse in a garbage bag would roll past the Death Row dogs, slowly, like a roller coaster car climbing to its first peak.

I spent a minute standing there, trying to figure out if the dogs understood their fate. Did they realize what was in those bags? Did they know they would be next? French philosopher Paul Valéry once wrote that "animals, which do nothing useless, don't contemplate death." I don't know. I doubt they *contemplate* death, but maybe they know it when they see it?

Somehow, I'd managed to keep it together until that moment, but I could only fake that I was okay for so long. I felt sick to my stomach. I wanted to vomit; I wanted to cry. I wasn't sure which one — tears, vomit — might come out of my body first. *Keep it together,* I told myself. *You're a professional.* I felt my knees buckle. Was I going to faint? *Keep it together. Act like you're in control of yourself.*

I took a deep breath. I looked at the dogs.

I could save them all, I thought to myself. I could open their crates, and they could make a run for it. Or, better yet, I could pack these dogs with me in the RV. I could be their savior. I could throw myself in front of their crates and refuse to move, a one-man protest of the heartless killing of dogs whose only crime was weighing more than thirty-five pounds on that late March afternoon in San Antonio.

But I didn't know anything about these dogs. They looked harmless in their crates, waving their paws meekly at me. Would it be practical to take them with me? Could I just take one? If so, how would I choose? What if I picked the wrong dog? What if the dog didn't get along with Casey? Would I then have to surrender the dog to another shelter? And how much time would Animal Care Services allow me to decide? They had a schedule to keep, more dogs to kill. What if I got Ellen's and Lisa's hopes up and then changed my mind, decided I couldn't take a dog? After all, I had a trip to finish, a book to write, a dog of my own to love.

I heard myself asking Lisa if we could leave. She said we could. I abandoned the dogs and staggered out of the EBI. I made my way toward the front, where the mangy owner-surrender was still resting on the floor. Holly and a coworker sat next to the dog, petting her. They couldn't save her, could they? Could they make room? An exception? It was a big dog — more than thirty-five pounds — and if they saved her they wouldn't be able to save another who was more likely to get adopted. "I don't like playing God," Holly had told me earlier. I'd believed her. I still did.

The dog — beautiful, mangy, sick, tired, in a battered brown collar — turned her head toward Holly and her coworker. Neither of them could help it; they looked straight into the dog's brown eyes. Holly's coworker began to cry.

"Okay," Holly said. "We'll take her."

On our second day in Marfa, Casey and I met a dog that almost made me forget about San Antonio. His name was Dersu, and at first glance he seemed out of place in the high desert of West Texas. A Malamute/ Husky, he had a heavy black and white coat, ears that stood straight up, and the kind of mystical, mismatched eyes — one brown, one blue — that seemed anything but accidental. The twelve-year-old dog was one of the most beautiful creatures I'd ever seen.

To my delight, Dersu was paired with humans who shared my reverence for him.

Nick Terry and Maryam Amiryani — both attractive, dark-haired painters in their forties — had moved from New York City to Marfa a decade prior, in search of a quieter, simpler existence. They'd bought a three-acre property on the far east end of town, with a backyard that overlooks an abandoned pecan orchard.

I'd learned of them through a woman at the Chinati Foundation (a contemporary art museum in Marfa) who had taken it upon herself to introduce me to some of Marfa's many dog lovers. Thanks to her, I found myself drinking green tea on the shaded patio of Nick and Maryam's wood-framed painting studios. As we spoke, Casey and Dersu napped on the grass.

"I just feel really privileged to live every day with this creature," Maryam told me.

"Thanks," Nick deadpanned, before breaking into a grin. "Oh, you meant *the dog*?"

Nick shared Maryam's admiration for what Dersu has added to their lives. "He tests my cynicism," Nick explained. "We have this idyllic life here in so many ways, but because we're human, we can get easily frustrated by things and lose a sense of what we have, of what the present moment offers. But for Dersu, each walk is truly thrilling. It makes his day! He sniffs the same spots. He walks the same forty-minute route twice a day every day. And he finds the beauty in it, the magic in it. In some ways it's absurd to compare a dog's life to a human life, but in another way it isn't. As humans, can we appreciate each moment, each day? I learn so much from him."

"Dersu teaches me patience," Maryam added. "He has infinite patience."

Patience had served Dersu well in the years he spent tied to a tree outside a trailer in town. Nick and Maryam had first spotted the dog six years earlier, in the spot where Dersu's previous owner would leave him every day with a bowl of water. He was underweight and his hair was matted, but Maryam and Nick noticed something else about him, too.

"He had such a beautiful demeanor and spirit despite his obviously terrible circumstances," Maryam recalled.

When they heard that the owner wanted to sell the dog, Maryam lied and told him she'd always wanted a Malamute/Husky. Two hundred dollars later, the dog was theirs. But they weren't sure what to do with him. "At first we didn't even plan to keep Dersu — we mostly just wanted to *free* him," Nick said. "We thought maybe we'd send him to Alaska. But we brought him home to see how things would go."

They went badly. Every chance he got, Dersu would jump their fence. "He was like a puma," Nick told me. Dersu would visit a female dog he liked across town or make his way back to the trailer and the tree. "It was like he was in an abusive relationship, like he had Stockholm syndrome — he couldn't stay away from this place where he was mistreated. When he was here, you could sense him trying to figure us out. He was like, 'I don't know about you guys.' And we were thinking, 'We don't know about *you*.' "

Nick worked at the Chinati Foundation back then, and when one of his interns planned to drive to Alaska, Nick offered to pay to send the dog with her. But one of the couple's dog-loving friends in town — just about everyone they know in Marfa has a dog — urged them to reconsider. And, before long, Dersu had settled in.

"Everything changed the day we offered him a piece of cheese," Nick told me with a laugh. "He was like, 'They have cheese, they love me, I'm staying.' He broke up with his girlfriend and stopped jumping the fence. And we stopped worrying that he needed to be in Alaska. He would shed his coat at the right time, and he would find places in the shade to keep himself cool. He's a Southwestern dog, born and bred."

In 2011 the couple founded the Dersu Collective, a group of Marfa artists who work to raise funds for community projects aimed at helping area youth. They showed me their first successful fundraising effort — a playground near their house, where Dersu gently leaned in to Casey to get a good sniff of his face. I noticed a bald spot on Dersu's left side; Nick told me he'd been bitten by a brown recluse spider, necessitating a "blitzkrieg" of medication.

"Dersu has such a high tolerance for pain, but he's also such a gentle giant," Maryam said. "His gentleness and friendliness, and simply the way he looks, helped break the ice between us and the neighborhood kids. People just gravitate toward his energy. The neighborhood kids look at him as their mascot."

Even Nick's mom, a purebred dog lover who'd warned her son not to adopt Dersu, ended up falling in love with the animal. "My whole life my mom would tell me, 'Never get a stray dog — they're a disaster,' " Nick said. "But when we went to visit my folks in California, Dersu was perfectly charming and well behaved, while my mom's precious purebreds were completely unwelcoming and really rude. She was mortified. It was a small victory for mixed breeds and mutts the world over."

After enjoying lunch next to train tracks at Food Shark, a Marfa staple widely considered one of the best food trucks in the country, I drove the RV to the west end of town to meet Alex Leos, Marfa's eighty-one-year-old former dogcatcher.

It had taken some legwork to find him. Though practically everyone in town knew of Alex, no one seemed to have his telephone number. Finally, a woman suggested I call or drop by City Hall and "speak to Lori," who apparently was the keeper of the town's telephone numbers.

When I arrived at Alex's modest red house on a corner, he was standing next to his two little dogs — a Chihuahua and a Jack Russell mix — in his cluttered, half-paved front yard. A late-afternoon storm was fast approaching, and the dogs paced nervously while wind gusts flung Alex's wind chimes practically horizontal. Alex, who has bright gray hair and thick laugh lines, wore gray dress pants and a short-sleeved collared mechanic's shirt, a vestige from his many years running a convenience store and gas

station in town. He'd retired in his late sixties but then took a part-time job as the
town's dogcatcher, puttering around in a
1970 Ford pickup.

"That clunker made so much noise, the
dogs would hear me from blocks away and
would go and hide!" Alex told me with a
hearty laugh. Still, Alex caught plenty of
roaming dogs in his eight years as Marfa's
dogcatcher. "A lot of folks in town would
let their dogs roam, which you're not supposed to do because there's a leash law.
They'd say things like, 'Here comes that
damn dogcatcher!' But other times they
were happy to hear that I'd picked up their
dog, because they were worried that it had
run away. Or they'd call me to help get rid
of a pesky skunk, or raccoon."

"You catch any feral cats?" I asked.

"Yes, and they were *mean*," he said. "If
you got a mean cat, you had better watch
yourself."

The town didn't have an actual shelter
when Alex took the job, so instead he used
a small shed that could accommodate only
a handful of dogs. On Saturdays, volunteers
would set up shop outside the courthouse
and post office and try to get the dogs
adopted quickly. If no one wanted a particular animal (or if it was "sick or mean," Alex

said), it would be euthanized. "It just had to be done," he told me.

Soon after, the town built an actual shelter near the Marfa Municipal Golf Course, the highest golf course in the state. And it was in that shelter that Alex, a lifelong lottery player, initially stashed his winning Texas Two Step ticket for $1.1 million. (His wife, Lola, then put it under their mattress at home before they moved it to a safety deposit box.) When Alex called the state Lottery Commission, they told him not to tell anyone that he'd won until he could fly to their Austin office.

"So I had to keep it a secret," he said, "and that was hard because everyone knew the ticket had been sold in town. People kept asking, 'Who won? Who won?' I just went about my business." Some townspeople suspected Alex, but he had an effective rejoinder. "I'd say, 'Do you think I'd still be driving this truck around catching dogs and raccoons if I'd won the lottery?' " Two reporters from the local paper still weren't convinced, and they tried to trick Alex into coming to their office so they could grill him by claiming there was a loose dog out front.

Winning the lottery could not have come at a better time for Alex and Lola. They'd been in debt, and Lola's health took a turn for the worse soon after. "Right after I won, she got sick," Alex told me. "She's diabetic, and her kidneys are gone. The good Lord was good enough to give me the money to take care of her until now. I kept saying to her, 'Now I can take care of you!' "

As we spoke in the living room, Lola sat in the adjoining kitchen without making a sound. She looked tired and ill; I couldn't tell if she was listening, or if she was lost in another world.

"What would have happened if you hadn't won the lottery?" I asked Alex.

He looked at me with his gentle, affecting eyes. "I don't know," he said. "I think about that a lot. Medicare doesn't cover everything, and without this money sometimes I think we would both be dead now."

Before leaving, Alex asked me if I'd bought a ticket for that night's Mega Millions Lottery drawing. I told him I hadn't. "If you want to win," he said, "you've got to buy the ticket."

"You still play the lottery?"

"Of course," he told me. "Who knows? I might even win big again!"

The next morning, I packed up the Chalet and headed north toward Truth or Consequences, New Mexico. On the way, I busied myself the same way I had for many long stretches of southern highway: I called Marc.

This time, though, I had some embarrassing news to share. "I think there's bed bugs in this RV," I said, mortified. I'd woken up each morning the previous week with itchy red bumps on my thighs. At first I'd assumed they were spider or mosquito bites, but they kept multiplying — and getting itchier.

On a late night at an RV park in Texas, I'd made the mistake of Googling "bed bugs in motorhome," which had pointed me to an article (titled "Bed Bugs: Old Problem Resurfaces in RV Parks") published on rv-business.com only two weeks after I'd begun my journey.

Marc was in his second year of medical school at the time, and he proposed several alternate theories for my symptoms. The first was scabies, contagious tiny mites that burrow into a person's skin. The second, and more preferable, possibility was folliculitis, a condition where hair follicles get inflamed and cause itchiness. I looked it up and learned that there's a variation of the ailment called "hot-tub folliculitis," caused by sitting in an unclean hot tub.

"I've been in hot tubs lately!" I told Marc, recalling late-night dips in several RV park tubs.

Still, I convinced myself the culprit was bed bugs. Though I'd found nothing when inspecting the mattress, there was plenty of dark, cavernous space I couldn't reach under the sleeping area. I figured they were hiding down there. I thought about all the people who'd slept on this mattress before me; an RV isn't quite a New York City hotel room, but it gets its fair share of traffic.

Wouldn't I have started itching sooner, though, if the bugs had been in the motorhome when I'd selected it in New Jersey? I wondered if I'd picked them up at an RV park along the way, or if Sam — or, worse yet, Marc — had inadvertently brought them into the Chalet in their luggage. And what about Casey? Could some bed bugs have hitched a ride on his fur into the motorhome? Can dogs transmit bed bugs to humans? I Googled the question and learned that I "shouldn't disregard the possibility."

All I knew for sure was that the whole thing was stressing me out. I had another seven weeks to go in this RV, and in three days my friend Garrett was supposed to join me on the road for a week, to film parts of my journey. Would he still come if he knew he might be taking bed bugs home with him? And what about Marc? We'd agreed that he would come visit again once I arrived in San Francisco. Would he change his mind? Would the second half of my journey turn into an itchy, traumatic nightmare?

Casey and I rolled into Truth or Consequences on a Thursday afternoon. Long known for its mineral-rich waters, this small city between Las Cruces and Albuquerque got its unconventional name in 1950 when the host of the popular quiz show *Truth or Consequences* promised to air the program from the first town that took its name. Hoping for a boost in tourism, the city of Hot Springs became the city of Truth or Consequences.

To that point on my journey, I'd fallen hard for two places — Savannah and Austin. I'd added each to the short list of American cities where I might want to live before I die. To my surprise, I was equally taken with Truth or Consequences (locals call it T or C). It had a sleepy, laid-back, mystical vibe — the Southwest at its best. Best of all, despite its attention-grabbing name, Truth or Consequences wasn't overrun with tourists. The few I did meet were as enchanted with the place as I was.

During a long, meditative walk with Casey at sunset along the Rio Grande, I bumped into three people — all locals who'd moved to T or C from big cities — walking their dogs in a group. They were curious about what I was up to, so naturally I told them about the trip, the book, the Facebook page. One woman said that she and her husband had also driven around the country in a motorhome with their dog.

"Of all the places we saw," she said, "we decided to move here. It's the most magical place."

Later that night, I took a soak in the Riverbend Hot Springs, an open-air spa perched along the Rio Grande. Three friendly college girls from New Mexico State University joined me in one of the pools. They'd brought wine, cheese, and grapes, and for the next hour we laughed together, mist rising from the bubbling water toward the exquisite, black night sky.

Day or night, the New Mexico sky can make poets of illiterates. Willa Cather described it in her novel *Death Comes for the Archbishop:* "The sky was as full of motion and change as the desert beneath it was monotonous and still — and there was so much sky, more than at sea, more than anywhere else in the world. . . . Elsewhere the sky is the roof of the world; but here the earth was the floor of the sky . . . the world one actually lived in, was the sky, the sky!"

It's a testament to New Mexico's beauty that I practically forgot about the parasitic insects likely multiplying in my motorhome. But I remembered them the next day when I woke up itchy — and with a hangover. I looked online for bed bug extermination companies, found one in Santa Fe, and made an appointment for that afternoon. I grudgingly said good-bye to T or C and drove north, past the small cities of Socorro and Belen and the village of Los Lunas (former home of rock 'n' roll legend Bo Diddley). In Albuquerque I connected to the Turquoise Trail National Scenic Byway, which took me to Santa Fe through rolling hills and old mining towns.

At the Santa Fe KOA, I spent three hours feeding quarters to the park's washing machines and dryers. Before meeting the exterminator, every linen and piece of clothing in the RV had to be washed and dried on high heat to kill the bugs. I stuffed items that weren't supposed to be dried in garbage bags and left them to bake in the sun outside the motorhome.

With my clothes clean (and sealed in

bags), I drove across town to the exterminator's office in a nondescript office park. I'd considered having the exterminator come to me, but I worried about setting off a panic in the RV park.

My exterminator turned out to be tall, slow-moving, and chatty. To my dismay, he didn't arrive with a bed-bud-sniffing dog (dogs have been used to sniff out infestations, especially in New York City) and seemed to want to talk about everything — sports, politics, dogs, writing — except bed bugs. Eventually, he began nonchalantly spraying the cabinets and mattress with a powdery white substance. I hovered around him, poking my head into every nook and cranny of the RV, hinting that he should talk less and spray more.

He never did find any bugs, or even any evidence of bugs. And though he didn't tell me that I was an idiot for hiring him, I got the sense that's what he was thinking. By the time I handed him a check, I was pretty sure I'd overreacted.

The next morning, as I walked Casey through the KOA, we came upon a young woman in a purple robe cradling a bulging toiletry bag and listening to music on her headphones. I assumed she was on her way to the campground's shower facility. I didn't have Casey on his leash, and for whatever reason he found the woman worthy of further inspection. He galloped toward her, his head held high. The woman didn't see or hear Casey approaching, and he made it all the way to her side before she let out a horrified shriek. Her toiletry bag hit the dirt with a thud.

"Get that dog away from me!" she screamed, pivoting stiffly to locate me, the irresponsible dog owner.

I called Casey back to me and apologized profusely. "I'm so sorry," I said, as she fumbled to remove her headphones. "He's friendly and just wanted to say hello."

"You need to learn to control your dog," she said, wagging her finger at me. "Dogs bite."

"Casey doesn't bite," I assured her. "He might lick you to death. But he doesn't bite."

She shook her head dismissively. "*All* dogs bite."

I wondered if she really believed that. After all, Casey hadn't bitten her — even as she'd screamed and nearly dropped her toiletry bag on his head. He'd simply backed away, his head down and his tail between his legs. As the woman picked up her bag and walked off in a huff, I was sorry that we'd met this way. I would have liked to have a chance to speak with her about her cynophobia — her fear of dogs.

Those who study animal phobias have found that while more people are afraid of spiders or snakes than dogs, living with cynophobia is considerably more challenging — especially today, as dog-wielding humans appropriate more and more public places.

When I'd spoken to friends (and friends of friends) about living with a fear of dogs, they described a debilitating phobia that affects where they go and who they see. "For the longest time, I would never go to the park because I might come in contact with a dog there," Margo, a school nurse, told me. "I'd question everywhere I was invited. If there were a chance I'd meet a dog, I wouldn't leave the house."

Margo decided to face her fear only when she noticed her daughter mirroring it. "I didn't want her to have to live like that," she said. Margo and her husband decided to get a puppy (they named it Casey), and though Margo initially kept her distance, she warmed up to the dog after a week or two. Today, she's much less fearful when she sees a dog in public. "But I'll still never be a dog person," she told me.

Like many women who suffer from cyno-phobia (men are considerably less likely to be afraid of dogs), Margo can point to an early traumatic incident. When she was five, she fell and skinned her knees as a big dog chased her down a sidewalk. I heard similar chasing stories from others. Robyn, a law student, said a neighbor's German Shepherd followed her for several blocks while she jogged as a young teenager.

"But little dogs scare me now, too," she said. "They can creep up on you and then start barking their heads off. You have no idea if they're going to bite you or hump your leg!"

Robyn worries that her cynophobia will hamper her future. What if she ends up marrying a dog person? What if she can't go to a best friend's baby shower because a dog is there? A fear of dogs can seriously impact a person's social life — and good luck getting sympathy from friends or family.

"Most people just tell me to get over it, as if it's that easy," said Sashana, a recent college grad. She hates when coworkers bring their dogs to work. "No one bothers to ask if anyone's bothered by it."

Sashana is black, and I asked her if she believed the commonly held stereotype that African Americans are more afraid of dogs than white people. "I wish that were true," she replied, "because then I could go over to more of my friends' houses."

But sociologist Elijah Anderson did find some evidence of racial differences, at least among working-class whites and blacks. In his book *Streetwise,* about a diverse urban neighborhood in Philadelphia, he noticed that "many working-class blacks are easily intimidated by strange dogs, either on or off the leash." He found that "as a general rule, when blacks encounter whites with dogs in tow, they tense up and give them a wide berth, watching them closely."

Kevin Chapman, a clinical psychologist at the University of Louisville, noticed the same anxious behavior among many African Americans that Anderson found. Chapman also discovered that nobody had explicitly investigated the incidence of cynophobia in African American populations. So in 2008, he and several colleagues conducted the first of two studies looking at the prevalence of specific fears across racial groups.

Compared to non-Hispanic whites, they found that "African Americans in particular may endorse more fears and have higher rates of specific phobias" — particularly, of strange dogs. When we spoke, Chapman offered two possible reasons. First, many dogs in low-income urban areas are trained to be what he calls "you-better-stay-away-from-our-property" guards. Being wary of those dogs makes sense — many of them *are* scary. In addition, Chapman told me, there's "the historical notion of what dogs have represented for black folks in America." In the antebellum South, dogs were frequently used to capture escaped slaves (often by brutally mauling them), and during the civil rights era police dogs often attacked African Americans during marches or gatherings.

As Chapman and his colleagues wrote in their 2011 study, many African Americans were psychologically conditioned to fear dogs when the animals were used as tools of racial hostility toward the black community. That conditioned fear is transmittable through families, he explained, and has contributed hugely to a community-wide fear of canines.

But though it seems that African American history has fostered a fear of dogs among some blacks, cynophobia mostly affects people who are conditioned to fear dogs *and* are predisposed to anxiety. When coupled, Chapman explained, environmental conditioning and genetic predisposition are "powerful enough to make someone develop a significant or substantial clinical fear of anything." And people who are that afraid — who have what Chapman calls "a legit phobia of dogs" — don't discriminate between canines, regardless of how their fear is conditioned.

"It may begin with a Rottweiler or a Pit Bull, or something that is stereotypically trained to be vicious," Chapman told me. "But if you're conditioned to think they're dangerous, that fear gets generalized to, say, Shih Tzus and Chihuahuas."

Regardless of race, those seeking help for their canine phobia have several therapeutic options. The most effective is in vivo treatment, where a therapist walks a person through instructions of increasing difficulty with a heavily trained dog, from leading the animal on a leash to, in the case of a brave patient, putting a hand in the dog's mouth. But as Margo proved when she and her husband came home with a puppy, you don't always need a therapist to recover from cynophobia. Sometimes, you just need to hang around a friendly dog.

When I'd visited Dr. Gold in Manhattan, he told me that one of his previous dogs, Amos, was helpful for patients who were afraid of dogs. One woman had been so frightened by the prospect of Amos lunging at her in his office that Dr. Gold promised a year of free therapy if the dog so much as approached her. She left treatment sometime later "kissing Amos on the head."

In Northern New Mexico, the mountains seemed to come from nowhere. They rose up along the edges of an otherwise flat, dry landscape, which was broken only by the huge, fauxdobe casino complexes north of Santa Fe. The mountains themselves weren't very big, but they congregated around the horizon like a group of people huddled in conversation on the opposite side of a room.

The contrast of mountains and flat land was more impressive in Colorado, where I was headed with Garrett (whom I'd picked up at the Albuquerque airport) one morning in early April to see cattle dogs in action. The peaks in the distance were taller than in New Mexico, but the road stayed flat and even through a desert that seemed to grow a little more green every mile, until it wasn't a desert anymore. Eventually the slopes weren't so distant, and the rocky cliffs pressed right up against the roadside barriers, which wrapped around the turns that started to kink the road along Highway 50.

It was a beautiful, narrow road full of

bends and curves, and around each new corner was a view of snowy peaks in the distance. The mountains that weren't snow-capped were brown and purple, but they were so covered with aspen trees that you could almost be tricked into thinking they were giant green hills. Despite the summits that had bubbled up from the landscape along the way, the asphalt remained smooth and straight; now it just seemed closer to the sky.

We were making our way to Gunnison, Colorado, a small city at the bottom of several valleys in the Rocky Mountains. Gunnison is ninety miles east of Ridgway, a town with an unforgettable unofficial welcome sign posted on the property of a disgruntled citizen: *Welcome to Ridgway. What we lack in wineries, we make up for in whiners.*

We'd been invited to spend two days in Gunnison with Rob and Bruce, a young couple. Along with Bruce's parents, they herd 1,200 head of cattle across 44,000 acres of rangeland each year with the help of their eight dogs, most of them Border Collies.

Garrett and I joked that we were headed to a real-life version of *Brokeback Mountain.*

I'd asked Garrett to join me because of

his skill behind the camera; I wanted him to document visually a portion of my journey. (I'd been taking pictures and video along the way with my iPhone, but he was bringing a real camera.) I'd met Garrett a couple years earlier in New Hampshire, where he was studying visual media and where I was on a two-month stint at the MacDowell Colony, an artists' retreat. When I told Garrett in passing that I was looking for someone to join me for a week and shoot video, he nominated himself.

Garrett loves coffee, and on his first morning with me in the RV he was shocked to learn that I didn't have a coffeemaker. He tried to play it cool, but I could tell he found me uncivilized. "Let's just stop and buy one as soon as we can," he said, annoyed and groggy-eyed.

We pulled into Gunnison late on a Monday night and found our way to Rob and Bruce's mid-century ranch house nestled amid cottonwoods and aspens and surrounded by hundreds of acres of hay meadow. As we gathered some belongings from the Chalet, Garrett confessed that he was nervous about spending two nights with guys neither of us had ever met. (Rob had read an article online about my journey and emailed an invitation to visit.)

Rob and Bruce were waiting for us when we piled out of the RV, and the second Garrett saw them relief spread across his face. Our hosts looked friendly and harmless; it probably didn't hurt that they also looked like J. Crew models. Both wore plaid shirts and stylish jeans, and they somehow managed — even when we'd see them the next day in mud-covered boots and gloves — always to look clean and put-together. Rob, the more outgoing of the pair, was tall and blond and looked younger than his thirty years. Bruce was twenty-seven, dark-haired, and more conventionally handsome. They were, Garrett and I imagined, the coolest cowboys in Colorado.

Rob told us that he never imagined living and working on a ranch. He had gone to school for journalism and was living in Denver when he met Bruce, who had grown up in Gunnison but wasn't sure he wanted to be a rancher like his parents. Bruce worried about being gay and alone "and having no life outside of ranch work." But when Bruce decided he was ready to return to Gunnison and his parents' ranch, Rob came with him.

Garrett — who surprised me with his eagerness to ask questions, most of them thoughtful and perceptive — asked the couple if they could see themselves living on this ranch thirty years from now. Rob and Bruce looked at each other and smiled. Then Bruce grabbed Rob's knee.

"I really want to see Rob as a crotchety eighty-year-old trying to get on his horse," he said.

Our first night in Gunnison was a short one; after setting us up in guest bedrooms, our hosts went to bed early. It was calving season, which meant they hadn't gotten much sleep lately. Like parents of a newborn, Bruce and Rob would alternate who had to get up at various points in the night to check on the cows.

"Yesterday we had a calf try to come out butt first, so we had to get that sorted out," Rob told us before bed.

Though I slept in both mornings in Gunnison, Garrett was obsessed with the idea of capturing a calf birth on video and told Rob and Bruce to wake him up if there was any action. But Garrett never got there in time. "Those babies are like ninjas in the night," he told me, shaking his head.

On our second day on the ranch, Rob and Bruce wanted to show us their Border Collies in action. We'd already met the dogs; they would meander to where we were staying from Bruce's parents' house down the road. At Rob and Bruce's, the dogs some-times hung out in the mudroom.

"These dogs work better when they don't live inside, but sometimes I'll *sneak them in the kitchen,*" Rob told me with a whisper, feigning guilt.

"Why do they work better when they live outside?" I asked.

"Well, working animals really shouldn't be pampered," he explained. "You care about them and love them, but they need to clearly know what their job is. If a pet gets humanized, that can confuse the dog about who is in charge and what's expected of her."

I heard something similar from another rancher, Marcia Barinaga, who uses several Great Pyrenees to guard her sheep in North-ern California. Normally Marcia doesn't socialize her guard dogs, but one, named Gordie, had consistent contact with humans as a puppy. "I may be reading too much into it," Marcia told me, "but I almost feel like he's having an existential crisis because he doesn't really know if he's supposed to be with the humans, or with the sheep. My other dogs know very clearly what their job is. They're very much professionals. But Gordie isn't so sure what he's meant to be doing, and sometimes he doesn't seem as content as the others."

The oldest of Rob and Bruce's dogs, a twelve-year-old black and white Border Collie named Tippy, was allowed inside their home the most. "She's worked so hard for so many years, she's earned her stay in the kitchen," Rob told Garrett and me. Bruce's parents often let her inside their place, too, where she likes to hang out next to an old steam radiator. "She loves to be hot. Even in the summer if we're driving in the truck, she's like, 'Roll up the windows, please!' She also loves sleeping in that truck. She slept in there the other night because we forgot her."

"Phew," I said. "I'm glad I'm not the only person who's forgotten my dog somewhere."

"Where did you forget Casey?" Garrett wanted to know.

I sheepishly confessed to having once left him in the yard at a friend's barbecue. "I was halfway home before I realized my mistake," I told them. "I called the party's host, and all I could hear in the background was everyone laughing at me."

"Oh, that's nothing," Rob said. "My mom once picked up the wrong dog from the groomer's."

"Seriously?" Garrett said. "How does that happen?"

"I guess she was having a really busy day, trying to juggle three kids and a dog and trying to get them everywhere they needed to be," he told us. "She'd picked up our Miniature Schnauzer, Penny, from the groomer and came straight to our elementary school to get my little sister and me, as well as my friend who was coming over that day. When we got into the back of our minivan, the dog snapped at my friend. Penny didn't normally do that, but my friend could be difficult, so my mom probably thought he had it coming. When we got home, the dog went straight into the living room and peed on the carpet. That's when my mom took a good look at the dog and realized it wasn't Penny. She'd brought home the wrong dog."

Though Rob and Bruce said that Tippy was happiest herding cows, they'd had to limit her activity in her old age. "She's got a bad shoulder," Rob explained, "and she just gets really sore if she's out herding too long. She's also kind of fat, which is my fault." He laughed. "I'm definitely the enabler, sneaking her treats. Bruce's mom keeps hoping we'll give her grandkids, and she knows I'll be the pushover dad."

On a frosty morning, Rob and Bruce

drove us about fifteen miles away to their upper ranch where their yearling steers were out to pasture. Tippy and another Border Collie, Fooler, were in the truck's rear bed, and I could see Tippy's expectant face in the driver-side rear window.

"The dogs know they're going to work," Rob assured us.

Dogs have been used for farm work since the Roman Empire, when sheep tending was introduced to the British Isles. Early collies — a name said to come from the Gaelic word for "useful" — exhibited the natural drive and skill to meet a variety of Celtic farmers' needs, Christine Renna writes in her book, *Herding Dogs*. The dogs would herd flocks of sheep, scout strays, hound cattle across great distances, and drive whole herds to market.

Farmers relied heavily on their working dogs, but it wasn't until the 1890s that breeders began intervening in the previously unrestricted mating of these sheepdogs, hoping to select for a handful of desirable skills. (Though effective herders, the dogs were often unfriendly, untenable, and rough with the sheep.) The result is the modern Border Collie, a "canine workaholic" with an endless drive to chase, circle, and protect a flock. They have no patience for easy tasks or busywork — collies are extraordinarily bright, and they've been known to get neurotic without mental stimulation. Stanley Coren labeled the breed the smartest in his book *The Intelligence of Dogs.*

At their upper ranch, Rob and Bruce showed us how the dogs push the cattle across seemingly endless acres of rangeland. Using the commands "by" to move the dogs to the left and "way" to move them to the right, Rob and Bruce directed the dogs — who moved with their heads low to the ground — to turn the herd or bring back cattle that had strayed too far. When Tippy ran to the left side of the herd, for example, the cattle would veer right in hopes of putting space between themselves and the dog.

Out on the range, in these rugged mountains only ten miles from the Continental Divide, the dogs are particularly useful pushing cattle out from among the willows or trees, or retrieving cattle from high mountain benches during the fall gather.

"It's a whole lot easier to move cows with a dog," Rob told me, "because they move a lot faster and can get to places you can't. We would spend way more time trying to manage and move the cattle without our dogs."

But dogs offer ranchers and farmers more

than just physical utility. Jon Katz, who has written several books about the animals on his Bedlam Farm, told me that dogs "provide farmers a sense of connection in a somewhat fragmented world. Farming is lonely and isolating, so it can be very important to have a dog."

Donald McCaig, a writer and sheepdog trainer who loves Border Collies, waxed poetic about his dogs in a piece for *The Bark* magazine, where he followed up a list of "practical" reasons for using sheepdogs with less tangible ones: "From the start, you'll have glimpses; momentary communication so intense, your and your dog's mind will be one mind."

I'd heard practically the same words from sled dog racer Lance Mackey, whom I visited in 2009 outside Fairbanks, Alaska, where he lived with his wife and teenage son in a house with no electricity and ten dogs. Mackey was talking about his longtime "lead dog," a fifty-pound Alaskan Husky named Larry, who had helped Lance win three Iditarods, considered the most grueling sled dog race in the world (it spans more than a thousand miles from Anchorage to Nome).

At the time of my visit Lance owned

eighty-five sled dogs, all of whom lived in small individual doghouses behind his house, where they would pace and howl in excitement at feeding time. Lance's sled dogs are very much working dogs, but the reverence he held for them was evident. Lance can be gruff and inconsiderate toward humans (he spent a good portion of my visit ignoring me), but he spoke admiringly about his dogs — especially Larry.

"We breed and train hundreds of dogs trying to find another one like him," Lance told me.

He compared his racing dogs to a football team. They all compete for only sixteen "starting" spots on his sled team, he explained, and if the more experienced dogs aren't careful, an eager younger one will bump him off the squad.

More than physical gifts, Lance says that what makes a great lead dog is a lot like what makes a great quarterback. "They have leadership that you can't coach," he said. "They either have it, or they don't. Larry has it."

Lance recalled one race in which Larry tried to lead the sled team in one direction, while Lance kept commanding the dog to take them in another. "As usual," Mackey told me, "Larry was right. I had missed the trail, but Larry knew exactly where we were supposed to go. How does he always know? I have no idea. But he's saved me from myself more times than I can remember."

7.
IN WHICH CASEY GETS
A LITTLE SISTER

It was dumb luck that I pulled the motor-home over for gas in Kaibeto (sometimes spelled "Kaibito"), an isolated community of 1,500 in Northern Arizona on the Navajo Nation Indian reservation. Had I not stopped the RV where I did, Casey and I

341

would never have met the dog that would change the course of our journey.

Garrett, Casey, and I had come that morning from the Four Corners Monument, where we ignored a "No Dogs Allowed" sign and stealthily photographed Casey lying on the quadripoint at the intersection of Utah, Colorado, New Mexico, and Arizona. From there we'd headed west toward Page, a small city near the border of Arizona and Utah where we planned to have dinner. But about halfway there, I had a craving for a Milky Way. As we rolled through the desert on Route 98, I stopped at the first gas station I saw: a Spirit, in Kaibeto.

The area around Kaibeto has been inhabited since at least 10,000 BC, marking the first traces of Anasazi culture in the Four Corners region. Some of the oldest archaeological evidence of the Anasazi comes from nearby Navajo Mountain. Kaibeto, which means "Spring in the Willows" in Navajo, was permanently settled in the 1840s and was formally recognized as its own Navajo Nation chapter in 1955. The Navajo Nation is the largest of the more than three hundred Indian reservations in the United States, covering some 27,000 square miles in three states (though it has a population of only

174,000).

Navajos live in some of the bleakest conditions in North America. Nearly half live below the poverty line, and many are sick — from addiction, diabetes, obesity, and, in a larger sense, historical trauma. That struggle extends to the hundreds of thousands of dogs who shuffle along the reservation's dirt roads, rest in the shade under its low-hanging trees, and beg for food at its gas stations and convenience stores.

"Rez dogs," as they're called, can be said to belong to everyone and no one. Most don't have a single owner; they're "community dogs," Kelsey Begaye, a former president of the Navajo Nation who lives in Kaibeto, told me.

Tamara Martin, the founder of the Blackhat Humane Society in St. Johns, Arizona, just off the reservation, added that "Native Americans don't have the luxury to care" about their epidemic of stray dogs. "There are so many other pressing issues, and there's an attitude of indifference toward the dogs that's born of helplessness," she said. "It's similar to all the garbage blowing around on the reservation. People would prefer it wasn't there, but nobody can pick it all up."

Lacking access to basic medical care,

many rez dogs die young from easily treatable illnesses. Some freeze to death on cold nights. On the Pine Ridge Indian Reservation in South Dakota, considered the poorest place in America, there isn't a single veterinarian. But dogs don't die only of neglect on reservations. Many are run over by cars — sometimes intentionally.

"Dog and animal abuse has increased among our people since the old days," Begaye told me. A tribal elder I would later meet on another reservation was more blunt. "Some of these teenagers are fools, cowards," he said angrily. "The younger generation thinks it's funny to run over dogs. They have no respect for the role these sacred animals play in our history."

Native American peoples have held fascinating and complicated attitudes toward dogs. Many tribes kept the animals as hunters, herders, fishers, long-distance haulers, and companions, while some also made use of them for food or ritual purposes. The Choctaw nation were among several Native American peoples who saw the human-canine bond as so important that they would kill and bury a dog with a deceased tribe member to provide him with a companion in the afterlife.

The Iroquois tribes used to sacrifice a

white dog as part of their Midwinter Rites. Tribe members would round up a dog without any marks and kill it by strangulation (ensuring that no blood was spilled), then decorate it with colorful ribbons, paint, feathers, and beads, and burn it along with a basket of tobacco as an offering to the deity Tanyawhagiy, "Holder of the Heavens." (In contemporary Iroquois practice, the white dog has been replaced with a white basket.)

Some tribes consumed dog meat as part of political or religious ceremonies, while others looked down on those who did. According to a belief common among Plains Indians, eating the meat was evil but conferred supernatural powers; by experiencing evil in small doses, one could build up a resistance to it. Unlike tribes who partook of dog meat only for ceremonial purposes, the Arapaho enjoyed it as part of a regular meal — so much that they became known among other tribes as the "Dog Eaters."

Other Native Americans valued dogs more for their hunting and working ability than their spiritual powers. Several Plains tribes, including the Shoshone and Kutenai, used dogs' chasing and tracking skills to help them hunt deer and elk. More commonly, tribes used dogs not as hunters but as carriers. Before tribes relied on horses, which came with the Spaniards, they used dogs to transport firewood, meat, and sometimes babies. Even after the arrival of horses (which the Lakota referred to as "sacred dogs"), the Hidatsa tribe preferred using canines to carry their loads.

But of all Native American tribes, the Yurok people — who live near the Klamath River along the Northern California coast — had perhaps the most interesting relationship to dogs. Though the Yurok valued dogs highly and treated them with great respect, the tribe never named or spoke to dogs. James Serpell notes that tribe members worried that the dogs "might answer back, thus upsetting the natural order and provoking general catastrophe." The Yuroks viewed dogs as especially dangerous because "the critical psychological line that distinguished humans from animals was constantly in danger of being effaced by their presence."

I knew nothing of this history — or even that we were on a reservation — when I swung open the Chalet's main door and stepped out into an overcast, windy early afternoon in Kaibeto. Clay dirt swirled over the cracked pavement of the Spirit Gas Station, which was advertising regular unleaded at under $4 a gallon, a rarity during that point of my journey.

As I yawned and stretched my arms toward the milky sky, I spotted a medium-sized black dog in an adjacent field of shrubbery and red clay. Though I couldn't tell for sure, the dog appeared to be playfully chasing something.

"I'm going to go check it out," I told Garrett, leaving him to hold Casey by the leash with one hand and pump gas with the other.

But before I could get far, the black dog spotted me and began bounding in my direction. Then, from behind the gas station, three more dogs appeared. One was long and gray and looked a bit like the wolf-dogs I'd met in North Carolina. Another was plump and brindle. The third was white, with steel blue eyes. And then there was the squinty-eyed black dog with a wavy flat coat, who I could now see had engorged breasts and appeared almost red from playing in the clay. All four animals looked up at me and wagged their tails.

No one paid the dogs any mind. People strolled in and out of the station store as if a pack of dogs hanging out at a gas station was a common occurrence. The only person to acknowledge the animals was a driver who pulled into the station in a beat-up red Honda Accord and honked once at the brindle dog; it was standing in the spot nearest to the store's entrance.

"I'm going to go inside and ask who these dogs belong to," I told Garrett as he filmed the pack with his camera. (He'd had to shoo Casey inside the RV because the steely-eyed dog, who had a gash on his right rear leg, had approached Casey with questionable intentions.)

"Do you know anything about the dogs outside? Do they have owners?" I asked the Native American woman behind the cash register. By the uninterested look on her face, I realized I probably wasn't the first visitor to ask about the animals.

"They're strays," she said. "They just hang out."

I walked back to the motorhome and grabbed two bags of freeze-dried Stella & Chewy's patties leftover from Global Pet Expo in Florida. The dogs loved the food — and the attention. But they somehow knew not to be pushy; they'd learned, it seemed, to be *polite* beggars. They didn't jump or body each other out of the way. When I knelt down, they would play-bow or shuffle up to my side one by one and let me pet them.

Of the four dogs, the black one stayed closest to me. She looked a bit like a miniature black bear. Though I'm terrible at guessing dog breeds, she seemed like she might be part Border Collie, part Flat-Coated Retriever. "There might even be some Chow in her," Garrett said. It was difficult to make an educated guess, though, because she looked like a clay-covered blob of black fur with swollen breasts. I wondered how recently she'd given birth, and where she might have stashed her puppies.

Back in the RV, I called a nearby animal shelter, but the woman who answered said the shelter didn't have jurisdiction to rescue dogs from tribal land. (I couldn't find any number for animal control on the reservation. I would later learn that there are only five animal control officers for the entire Navajo Nation.) Then I called Patty Hegwood, my contact at Best Friends Animal Sanctuary in Kanab, Utah, where Garrett and I were scheduled to arrive the next day. Best Friends runs the largest animal sanctuary in the country, housing some 1,500 dogs on 3,700 acres.

"You won't believe what I stumbled on!" I told Patty when I reached her by phone. I sat on the RV's passenger-side steps as we spoke, with the door open so I could keep an eye on the dogs. "I pulled over for gas on the Navajo reservation, and there are four stray dogs here just hanging out at the gas station."

I could almost hear Patty smiling at me as I prattled on about my discovery. "Welcome to the world of rez dogs," she said finally. "Everyone at Best Friends remembers their first trip through the reservation. It's gut-wrenching. A lot of people I know will drive way out of their way to avoid the reservations, because it's just too painful. There's an endless amount of strays."

I told her about the black dog. "She's so sweet," I said. "If I bring her with me, would you, by chance, be able to take her at Best Friends?"

"It's possible, but we can't guarantee it," Patty told me. "If you take her, you should be prepared to keep her."

Keep her? As my own? In the Chalet? I wasn't so sure I could do that. Though the black dog seemed charming and harmless *now,* there was no telling how she might react to the RV. I knew what to expect from Casey: I could leave him alone in the motorhome without incident. But this black dog was a mystery. I worried that if I left her in the RV while I was away, I might come back to find the inside torn to pieces.

I remembered what Neil and Brant's rescue, Amelia, had done to their bedroom.

I also had no idea if she'd get along with Casey. I knew that Casey wouldn't mind *her* (he's easygoing and friendly with just about anything that moves), but did he really want a permanent little sister? Finally, did I have the time — or the energy — to devote to a new dog while driving around the country?

I changed the subject and told Patty about the black dog's engorged breasts. "It looks like she just had puppies," I said. "I wouldn't want to take her from here if she's nursing puppies."

Patty instructed me to squeeze one of the black dog's teats to see if any milk came out. I stood up and walked a few paces to the dog, who was lying on her side on the concrete, dust blowing into her face. I squeezed. Drops of milk dribbled onto the pavement.

"She probably had puppies fairly recently," Patty surmised.

"So what should I do?" I asked.

"It's hard for me to say not to rescue her, because it might be her only chance at a good life. And it's also hard for me to tell you to rescue her, because her puppies might die."

351

"That's not helpful at all," I said, laughing at the absurdity of my choices.

I thanked Patty and went searching for the puppies. "They can't be far," I assured Garrett, as we walked toward the back of the gas station. Garrett looked at me quizzically. "And what if we find them?" he asked. "Are we taking them all in the RV with us to Best Friends?" He didn't seem enamored of the idea of transporting peeing and pooping dogs.

I searched under Dumpsters, sheds, pickup trucks, and shrubs. Every once in a while I would look at the black dog and say, "Are your puppies over here?" "Over there?" "Underneath this shed?" She followed me around; maybe she thought I'd misplaced something. I asked a few people standing outside the station if they'd seen any puppies, but they hadn't. Finally, as the sun was setting over the horizon, I knocked on the front door of a trailer home next to the station's back parking lot. A young, groggy-eyed woman answered.

"I'm sorry to bother you," I said, "but do you happen to know this dog?" I pointed to the black animal standing at my feet. By that time, I'd brushed most of the clay from her fur. She looked slightly more presentable.

"Yeah, she's one of the strays around here," the woman told me. She was friendly and seemed eager to help. "She's a nice dog. She had puppies about a month or so ago."

Finally, a lead! "Do you know where they are?"

"They used to be under that," she said, pointing to a dumpster I'd already inspected. "But people took them."

"People took them?"

"Yeah, they like taking the puppies. No one really takes the older dogs, but people think puppies are cute."

"Is there anything else you can tell me about this dog?" I asked her.

"Well, she got hit by a car a while back."

"That explains the limp," I said. It was barely noticeable, but she favored her right leg.

"She's a real sweetheart," the woman assured me. "Are you thinking of taking her?"

I nodded. "It's the craziest thing in the world, because I'm driving around the country in an RV," I told her. "But I'm considering it."

Back in the motorhome, I called Marc. He listened patiently as I told him about Kaibeto, the gas station, the stray pack, and the dog I wanted to kidnap. I explained that there was no guarantee Best Friends could keep her.

"I might end up with a second dog," I told him. "Am I being incredibly reckless?"

Marc was careful not to make my decision for me, but I could tell he worried that a new dog came with risks. "I just know how much you've said this trip is about you and Casey bonding," he told me. "Rescuing a new dog doesn't sound like it will help you do that. And do you really want to take on the responsibility of a stray dog right now?"

But, I thought, rescuing this dog might help me atone for my purchase of Casey in 2003. I'm embarrassed to say that back then I knew nothing about puppy mills, which the ASPCA defines as "a large-scale commercial dog breeding operation where profit is given priority over the well-being of the dogs."

In my ignorance I'd traveled to a now closed pet shop in Hopkinton, Massachusetts, which I later learned allegedly sold dogs from mills. The pet shop was later investigated by the state after customers complained that their dogs suffered from "pneumonia, kennel cough, giardia and campylobacter, parasites that can spread to humans, and a host of other ailments," according to the *Boston Herald*. When people would ask me where I'd gotten Casey, I always tried to change the subject.

I decided to sleep on my decision about the black dog. I got permission from the gas station to park the motorhome overnight in its back lot, but Garrett wasn't thrilled about the idea. "I'm sure we'll be fine," I said. I pulled out a small kitchen knife from a drawer and flexed my biceps. "If anyone messes with us, I'll use this."

Garrett laughed. "You're ridiculous," he said. "You don't have mace? Or a gun?"

Garrett wasn't the first person to question my decision to travel around the country unarmed. Several friends had suggested I bring a gun, mace, bear spray — anything at all that might keep me alive in a dangerous situation. But I'd decided to put my faith in the good-heartedness of the American people. "You're an idiot," one especially paranoid friend told me before I left.

I wasn't sure what to do with the black dog overnight. I worried that if I didn't keep my eye on her, she might meander to another part of the reservation, never to be seen again. But I also didn't want her to spend eight hours in the RV without a thorough bath. We'd watched rez dogs itch themselves silly; Garrett figured they were carriers of fleas, ticks, and other bugs we hadn't yet heard of. I was probably more

355

paranoid about bugs than I should have been, given my bed bug scare.

I decided to tie the dog to the side of the RV with a rope. I felt guilty about this, but she didn't seem to mind. I checked on her every few hours, and each time she looked perfectly content. At one point I thought she'd gotten away, but then I peered under the RV and found the dog curled up next to a rear wheel.

"Good girl," I whispered. "We'll have you out of Kaibeto in no time."

I had apparently made up my mind.

The next morning, Garrett and I ripped apart several garbage bags in order to create a makeshift tarp for the dog to lie on during the two-hour drive to Best Friends. She was leaking milk and didn't seem very comfortable in the motorhome, so I wrapped her in the one ThunderShirt I had left and asked Garrett to sit next to her and comfort her while I drove.

I don't remember the scenery on the way to Kanab; I was too preoccupied with our new passenger. Her body trembled slightly at first, but she eventually calmed down and appeared to go to sleep. "She really is cute," Garrett said, petting her on the head.

We arrived at noon in Kanab, a small city

not far from the Grand Canyon's north rim. It was founded in the 1850s by Mormon settlers, though it's perhaps best known as "Little Hollywood" because of the many westerns — including *Gunsmoke* and *The Lone Ranger* — filmed there over the years. The Best Friends Animal Society moved to town in 1984 and sits in Angel Canyon, a dramatic red rock canyon carved by Kanab Creek.

Garrett and I parked the RV by the sanctuary's veterinary clinic, where Patty had agreed to have the dog checked out. But the second we reached the clinic's front door, the dog stopped. Either she'd never been indoors before or the experience had been unpleasant enough for her not to want to repeat it. We tried all manner of sweet talk to get her inside, but the dog wouldn't budge until a staff member laid out a trail of treats on the floor. Still, each step was labored and cautious. I got down with her on all fours to encourage the process, but after a minute or two a staff member swooped in, lifted her into his arms, and disappeared with the dog into a back room.

We had a few hours to otherwise occupy ourselves while they examined and spayed the dog, so Best Friends' John Garcia — the star of *DogTown,* a National Geographic series that chronicled the lives of dogs and staff members at the sanctuary — took Garrett and me on a tour of the sanctuary. John has a soft spot for dogs who "get a bad rap," which might explain why he's so taken by Pit Bulls. Not long into our tour, he led us to a large outdoor kennel where a brown Pit wearing a red bandanna lounged in the dirt. The dog jumped up as we approached, his tail wagging as he stuck his nose against the fence. Cuts and scars crisscrossed his face.

"This is Lucas," John said. "He was Michael Vick's best fighting dog."

When Lucas and dozens of other Pit Bulls were rescued in 2007 from Vick's dogfighting compound, PETA and the Humane Society of the United States had recommended the animals be euthanized. But a judge ruled that twenty-two of the "most challenging" dogs be sent to Best Friends for rehabilitation. In Lucas's case, because of his supposed viciousness and his value to potential dogfighters, the judge ordered the dog to spend the rest of his life at the sanctuary.

Best Friends dubbed Vick's Pit Bulls the "Vicktory Dogs," and Lucas was perhaps the most outgoing and charismatic of them all. He adored people, and everyone at Best

Friends adored him back. Lucas and the other Vicktory Dogs received extensive media attention (they were even the subject of a book, *The Lost Dogs*), which John said helped improve the perception of Pit Bulls.

"For the longest time people assumed that fighting dogs were beyond hope, but we proved that they can be rehabilitated," John told us, adding that several of the dogs would soon be adopted into homes. "And every time someone meets Lucas or these other dogs, they realize that Pit Bulls aren't what they're made out to be."

Best Friends advocates against breed-specific legislation (BSL), which restricts or prohibits the ownership of certain breeds — namely, Pit Bulls — on the premise that they're inherently dangerous. In the last few decades, hundreds of municipalities across the country have adopted such bans, even though a University of Pennsylvania study found that Pit Bulls are about as likely as Poodles — and significantly less likely than Chihuahuas — to attack humans. (That said, a Pit Bull's bite can do significantly more damage than a Chihuahua's.)

The public perception of Pit Bulls has changed markedly since the 1980s, when the breed had a reputation for being charming and athletic, appearing in family films and as a mascot in various advertisements. But after a series of well-publicized maulings in 1987 among which a Pit Bull killed a toddler, the dogs' image was transformed into one of violence and aggression. It probably doesn't help that studies have linked Pit Bull ownership with criminality.

The first Pit Bull–specific ban was passed in 1987, and the breed has since been singled out disproportionately in BSL bans. But as critics point out, the public's perception of dangerous dogs tends to change over time. "Dobermans, Rottweilers, and German Shepherds have all gone through periods when they were seen as the ultimate 'tough' dog," James Hettinger wrote in *Animal Sheltering* magazine. (Other breeds have also been stigmatized. In 1738, a Blue Paul Terrier — a breed that no longer exists — attacked a number of other dogs and humans in an Edinburgh meat market. In response, enraged citizens converged on the dog, slaughtered it, and placed a one-pound bounty on all Blue Pauls. Within days, dog carcasses hung from shop windows, floated in the harbor, and lay on the side of the road, clubbed to death.)

Some sixteen states have banned breed-specific legislation in recent years, arguing that such laws are unfair and ineffective.

That's good news to John Garcia and Best Friends, who have made it part of their mission to show that dogs aren't born bad, and that most dogs made bad by humans can be rehabilitated.

After visiting the outdoor dog kennels, John took us for a tour of Angel Canyon in his Jeep. He drove us down a winding road, past a horse sanctuary to a cavernous lake under a canyon. "It's like a scene from *The Abyss*," Garrett said, adding that he half-expected some underwater ET to float up in a bioluminescent ship, or for a monstrous tentacle to pull us all under.

From there John took us to Angels Rest, a pet memorial park and the final resting place for many sanctuary animals. Low wooden poles, each adorned with a chandelier of metal wind chimes, had been erected around the flat, brush-lined yard. The chimes jangled in the breeze over rows of terra-cotta-colored gravestones, each of which lay flat against the gravel. Without headstones protruding from the ground, Angels Rest felt less like a somber graveyard and more like a well-tended Zen garden. I stood beneath one of the wind chime curtains in the center of several graves, which had been arranged in concentric half-circles, like the top half of the sun as it peeked out from a rocky red cliff behind the park.

When Garrett and I returned to the clinic, a young vet told us that the black dog was doing fine. The spay surgery had gone well. "But it's good you found her when you did," she said, adding that the dog would have died without medical attention.

She spoke matter-of-factly, explaining that the dog had a pyometra, a uterine infection that causes the uterus to fill up "like a water balloon full of pus," she said, cupping her hands around an imaginary expanding water balloon as she spoke. "It's a raging infection, and it happens very quickly."

"And it would have been fatal?" Garrett asked.

"Yes," she said. "It was an early-stage pyometra, so she probably had another week to ten days."

The next morning, we returned to the clinic to pick up the dog. Though they'd hinted that they could make space for her at the sanctuary, everyone I talked to at Best Friends seemed to be rooting for me to keep her. Still, I wanted to spend one more night in Kanab to see how she handled life in the Chalet. To Garrett's dismay, this extra time in Utah meant we'd have to skip the Grand

Canyon and drive directly to Phoenix, where he had to catch a flight back to New Hampshire.

Another vet met us in the clinic waiting room and proceeded to deliver more unexpected news. "I'm concerned she might have *Brucella canis*," she told us. *Brucella canis*. The words practically rolled off the tongue and sounded to me like some sort of harmless, perhaps even aristocratic, canine ailment.

"It's a sexually transmitted disease," she said. "It's a pretty common disease on the reservation."

The news got worse: *Brucella canis* can be transmitted to humans through body fluids, including saliva and milk. "Your dog's already potentially been exposed," she said, "and so have you."

It took a moment for the words to sink in. *My dog might have given me an STD.* I thought about the awkwardness of breaking the news to Marc. "Let me preface this conversation by saying that I'm not sexually attracted to animals," is how I imagined that might go. But there was good news, the vet said. "It's a low likelihood that she has it," she told us. "She's not showing all the signs, but we were concerned by some of the ways the tissue looked when we did the spay."

The vet handed me a bottle of doxycycline for the dog and suggested that I keep her away from other animals and people for two weeks. "Just to be safe," she assured me.

I worried about keeping the black dog away from people and dogs. What kind of start to her new life was *that*? Was I just going to isolate her in the motorhome, like some sort of leper? I wondered what I would tell my dad and stepmom when I visited them in Scottsdale. I'd been looking forward to a stress-free few days in their desert home. *Brucella canis* might just ruin everything.

John Garcia tried his best to cheer me up later that afternoon at an outdoor shooting range facing a cliff. A man of remarkable enthusiasm, he handed me a shotgun and a beer and ordered me to "have some fun." I wasn't so sure I should be drinking during my introduction to firearms, but I turned out to be a surprisingly good shot with a shotgun. Garrett was terrific with a handgun. John was probably just being nice, but he said we were two of the best first-timers he'd ever seen.

Back at the RV park that night, the dog mostly slept in a crate John gave us for the road. He'd warned us that she might be tired for a few days — the surgery, combined with the stress of a new environment, could make her seem lethargic and distant.

While she slept, I searched the Internet for information about my potential STD. On the website VeterinaryPartner.com, I found a report detailing the illness in dogs. It wasn't pleasant reading. Though antibiotics are used to treat *Brucella canis,* the article made clear that the "bacterium is so good at hiding inside the host's cells, you can never assume it is ever truly gone." Other articles suggested euthanasia as a common "treatment." I looked up the symptoms in dogs; they included inflammation of the spine, kidneys, and inner eye. But often the only sign of the illness is a miscarried pregnancy.

That would have seemed to rule out the black dog, who had apparently delivered living puppies. But the vet at Best Friends had cautioned me that the woman I spoke to on the reservation could have been mistaken; maybe those puppies had actually belonged to a different dog.

Next, I found a 2012 report by the National Association of State Public Health Veterinarians titled "Public Health Implications of Brucella canis Infections in Humans." The report found that there were between one hundred and two hundred cases of the illness in people reported each year, but that "it seems likely that *B. canis* infections in humans are significantly underdiagnosed and under-reported, primarily due to the nonspecific presentation of the disease and the lack of readily available laboratory testing." Fortunately, the treatment is easier in humans than dogs; normally a course of antibiotics does the trick.

Garrett, for his part, took the STD news in stride. "At least it's not worms or parasites," he said. "I'll take dog syphilis over worms any day." He could sense that I was making myself crazy online, so he temporarily broke the spell by announcing that it was time to name the new dog.

"We can't just keep calling her 'Black Dog,' " he said.

He was right. Now that I was pretty sure I was going to keep her, it was time to name the animal. But I wasn't feeling especially creative. The first idea that came to mind was "Bear," to honor the passing resemblance. "Too common," I said. We sat there for the next thirty minutes, drinking wine and coming up blank.

Just when I was about to give up for the night, I blurted out "Rez."

"Why not name her Rez? Or Rezzy?" I

proposed. I'd never met a dog named Rez before, and it fit with the common practice of naming a stray dog for where you found her. I suppose I could have also called her Spirit, after the gas station that was her former home, or Navajo. But Rez, or Rezzy, seemed to fit. I knelt down in front of the dog's open crate, where she was curled up in a ball.

"Nice to meet you, Rezzy," I said.

On our way south the next morning to Sedona, Arizona, I pulled the RV over at Lone Rock Beach on Lake Powell, a stunning reservoir on the border of Utah and Arizona. The second-largest man-made reservoir in the country (by capacity), Lake Powell has two thousand miles of shoreline, more than the west coast of the continental United States. Lone Rock Beach is named after a solitary rock that sits in Lake Powell's Wahweap Bay, some 750 yards from the beach.

Visitors can dry camp at the water's edge, and on the day of our arrival (Easter Sunday), there were a dozen or so all-terrain vehicles and campers parked on the muddy sand. I didn't want to chance getting stuck, so we left the RV in a lot and walked down to the beach with the dogs. Though Rezzy still seemed dazed from her uprooting and operation, I was curious how she'd respond to a body of water. I doubted she'd ever seen one.

As he is wont to do, Casey made himself at home on the beach. He swam far out into the water after a tennis ball and then returned to land to destroy a young boy's inflatable beach ball. The poor boy sobbed on his beach towel. I apologized profusely to his parents, who turned out to be good sports about it. "There are many more where that ball came from," the dad assured me.

With that crisis averted, I let Rezzy off the leash. Garrett and I watched as she walked gingerly along the water's edge. Though she didn't go far, her tail never stopped wagging. I was hoping she'd go to the bathroom, because she hadn't pooped since we'd picked her up the previous morning from the clinic. But no such luck.

After an hour on the beach, we climbed back in the motorhome and drove south to Sedona, a small city known for its scenic red sandstone formations and its spiritual, New Agey vibe. We spent that night at a dog-friendly hotel — El Portal — and had dinner and margaritas with the dogs on the patio of a Mexican restaurant. Though we didn't have much time the next day to explore the area, we took a short morning hike through the red rocks. I let Casey walk ahead of us but didn't let Rezzy off the leash this time. There were too many places where she could potentially run and hide.

Rezzy didn't poop on that walk, either. I called the Best Friends vet later that morning to ask for advice, but she said not to worry. "Sometimes the stress of a new environment can cause that," she assured me. "Give me a call if she still hasn't gone in forty-eight hours."

While I had her on the phone, I asked if the *Brucella canis* test results had come back yet. They hadn't. I pressed her on the likelihood that Rezzy had the illness. "It's unlikely," she said, repeating what she'd already told me. "But we should know for sure soon."

As Garrett and I drove toward North Scottsdale, where my dad lives in a beautiful red adobe desert house, I confessed that I felt like a terrible son. I hadn't told my dad and stepmom about Rezzy's possible condition, and I wasn't sure I was going to. "If I tell my dad," I said to Garrett, "he's going to freak out about getting it and will probably want me to keep Rezzy and Casey in the motorhome for four days."

"But it's too hot to keep them in there

during the day," Garrett said. "Maybe he'll just have you confine the dogs to one room in the house?"

That was a possibility. But it still sounded stressful. My dad doesn't react well to changes in his routine, and a newly rescued street dog with a potential STD that could be transferred to himself, his wife, and his dog would constitute a serious disruption to his peaceful quasi-retirement.

Besides, I was mostly convinced that Rezzy didn't have the illness. I'd believed the woman on the reservation when she told me that Rezzy had birthed healthy puppies, and from everything I'd read online, a dog with *Brucella canis* would have had a miscarriage. Still, I couldn't help being on edge my first day at my dad and stepmom's house. My body tensed when my stepmom, an animal lover who emails me several funny dog or cat videos each month, knelt down to kiss Rezzy. Later that day, during an evening walk with the dogs through the desert, I hip-checked Rezzy out of the way when I sensed that my stepmom was about to go in for another smooch.

"I'm going straight to hell," I told Marc when I spoke to him by phone that night.

Thankfully, Rezzy turned out to be fine. The vet called me the next day with the good news — my new dog didn't have "doggie syphilis." During a long walk in the desert to celebrate, Rezzy even decided to defecate; she did so behind a cactus, which offered her some semblance of privacy.

My dad and stepmom share their home with a beautiful, sensitive, tennis-ball-obsessed Golden named Cassie. The dog is a celebrity of sorts, at least in the world of Arizona Golden Retriever rescues; she was the 2011 Arizona Golden Retriever Connection calendar cover girl.

Like many cover girls that have come before her, Cassie can be high-maintenance. She's suspicious of some dogs, particularly those that compete for attention from her favorite humans. She seemed especially threatened by Rezzy, perhaps because they're both females. When my stepmom tried to play with the dogs on Cassie's favorite carpet, Cassie bit Rezzy on the top of her nose. Rezzy slinked away with a small cut.

Though Cassie has never bitten me, she wasn't my biggest fan when I'd first met her the previous year. In fact, she couldn't stand to be in the same room with me back then. If I got too close, she'd shuffle backward or pee on the floor. She'd quickly warmed up to my dad, but she would cower in fear at the sight of most men.

"I'm not going to give up without a fight," I'd assured her, lying on my back on the ground (the most unthreatening position I could think of) and making sweet kissing noises. Other times, I would throw tennis balls for her or toss chunks of salami in her dog bowl when I was sure she was looking.

In the midst of Cassie's obvious anxiety around me, I'd suggested to my dad and stepmom that we reach out for help to a pet psychic — or an "animal communicator," as most prefer to be called. I'd been reading about animal communicators as initial research for this book, and I'd even spoken to several about my relationship with Casey (more on that in a bit).

I searched online and found a local communicator by the name of Debbie Johnstone, who had been profiled two years earlier in *The Arizona Republic.* Though most animal communication sessions are done over the phone, since Debbie was local I called and made an appointment for her to come by the house and meet Cassie.

In person, Debbie managed to look both whimsically childlike and intensely serious. Her gaze was direct but not piercing; it was softened by oval glasses and bright blond bangs that curled slightly inward over her much darker eyebrows. Her hair was the almost-white color one might find on a storybook princess, but her face also had a slightly masculine edge. She looked a bit like actor James Lipton with a wig.

"I believe I was born this way," she told us, seated next to Cassie on the floor of my dad and stepmom's home. "When I was two, I remember talking to the animals and hearing them. My mother thought it was my imagination, but I knew I could communicate with them."

Debbie's childhood connection to animals is typical of the hundreds of animal communicators — the vast majority of whom are women — working today in the United States. Penelope Smith, who has been called the "grandmother of animal communication," credits her childhood cat, Fritzi, with comforting her while her parents fought.

"We took refuge in our bedroom to avoid the frequent parental conflicts accentuated by alcohol," she writes in her book *Animal Talk*. "Fritzi and I understood each other deeply."

Though the practice of animal communication existed before Penelope made it her job, the profession didn't gain traction until the late 1970s. The field's growth coincided with the burgeoning New Age movement; many communicators share an interest in alternative medicine and mystical philosophies.

Animal communicators insist that anyone can learn to commune with their pets, but "it takes practice," Debbie told me. For Debbie, that began soon after the September 11, 2001, attacks. She'd been working for nearly twenty-five years in the technology industry, but 9/11 shook her out of the complacency of living a life she wasn't passionate about.

"My job didn't feel right, and neither did I," she told us. "I decided to go do what I loved most — working with animals." She started volunteering at a rescue organization for horses and took a four-hour animal communication workshop. "Eventually my ability just turned back on," she said.

Many hopefuls attend workshops, where they say they use focused meditation to receive images and feelings from other species telepathically. Other communicators rely on alternative techniques: some claim to go into a shamanic trance and speak to an animal's "spirit guides," while Houston-based celebrity pet psychic Sonya Fitzpatrick says that animals project their thoughts and feelings through the earth's magnetic fields.

Animal communicators insist they act as translators, conceptualizing the messages they receive — which they say can come through words, pictures, or feelings — into affirmations of our animal companions' happiness or explanations of their misbehavior. Joan Ranquet, who teaches at Communication with All Life University, a school she founded, finds that dogs are a lot like hyperactive five-year-olds. She takes notes as she listens, which may read something like, "I'm a good dog. I love the smells in the yard. The woman was wrecked over her breakup. I'm a good dog."

(Animal communicator Shira Plotzker says that "dogs can talk about pretty much anything . . . cats are more particular about what they want to talk about, and it's mostly about themselves.")

Perhaps because of my conflicted relationship with Casey, I assumed that if animal communication was real (and I was skeptical about that), that pets might use their time on the therapist's couch to reveal shameful secrets about their humans. But our pets rarely betray us, according to most of the communicators I spoke to. In fact, Debbie told me that animals just want their humans to be happy.

"But what would you do if an animal told you that his owner was beating him?" I asked her. "What would be your responsibility there? To save the dog, or to report that to the human?"

"That's never happened," she said. "I can't imagine that someone who beats their animal would hire me to better understand their animal."

"It's an interesting ethical question, though," I insisted.

"Yeah, I might say to the animal, 'Okay, you sneak out the back and meet me in my car!' "

I was curious about what kind of people hire animal communicators. I was certainly an unusual client; I was writing a book about dogs. But Debbie assured me that I was typical in one important way — I was going to take some convincing that her skills were real.

"People are sometimes so desperate to understand what's going on with an animal either behaviorally or physically," she said, "that they'll throw up their hands and call me and say, 'I don't really believe in this, but I'm willing to try anything!' "

Many of those who hire communicators are looking for guidance about whether to put down an old or severely ill pet. Carol

Gurney, a well-known animal communicator based in Los Angeles, told me about several such cases, including that of a seventeen-year-old Lhasa Apso named Ozzie. The dog was incontinent, could barely move, and wasn't eating. Ozzie's vet had recommended that he be put to sleep, but the dog's owner, Valorie, wasn't sure.

"Valorie told me, 'We as a family are willing to let him go, but we want to be sure that's what he wants,' " Carol recalled. "So I asked Ozzie, and he said, 'No, I'm not done with my job.' " Believing that Ozzie could benefit from some healing bodywork, Carol went to Valorie's house. "Ozzie was on the bed and he looked almost dead. He couldn't move. I really questioned myself. How could this be true? How could he not be ready to go? I asked him again. He said, 'No, I'm not ready to go. I'm not done with my job.' "

Carol says she encouraged Valorie to continue the bodywork after she left, and three days later Valorie called to report a "miracle." Ozzie was running around the backyard like a dog half his age. "I almost didn't believe it," Carol told me. "I just figured we would make Ozzie comfortable, but he lived another year. And then I realized what Ozzie meant when he said he wasn't done with his job. Valorie had had an aneurism when she gave birth and was learning to walk all over again. I understood that Ozzie was her best friend, and he wasn't ready to leave until she was further along in her healing."

Carol and Debbie both told me that dogs understand death, but that they're much less fearful of it than we are. "Animals that are connected to a human just want to know that they're leaving us at an okay time," Debbie said. "They want to know that we'll be okay without them."

I didn't outright reject the possibility that animals might have important things to tell us, but my phone sessions with several communicators about my relationship with Casey had done little to diminish my skepticism. Our talks had been remarkably devoid of "verifiable information," which means that very little of what they said rang true.

One woman, in particular, lost me early in our conversation when she announced that Casey wanted me to "have a special girl in my life." Another said that Casey was a "fast and gliding runner," when the truth is he's slow and hobbles slightly. What wasn't outright wrong was usually absurd ("Casey really wants to be on the cover of your book!") or vague ("Casey likes to get along."

"Casey likes to play outside." "Casey thinks rules are hard but he really tries.").

Although I'd been mostly unimpressed with the communicators I spoke to, I didn't believe them to be con artists. I believed that *they* believed they could understand the yearnings of animals.

I could tell my dad was growing impatient when Debbie had been at the house for thirty minutes and we'd barely even mentioned Cassie. "What has Cassie been telling you so far, if anything?" I asked her.

Debbie was sandwiched on the floor between the couch and the coffee table, gently holding a tennis ball at the dog's eye level. Cassie sat at attention in front her, pawing at the ball. They appeared to be having a moment.

"She likes to divulge herself very slowly, so I'm not pushing her," Debbie told us. She looked at my dad and stepmom. "Does Cassie ever get to see her sister?" They shook their heads. "Cassie's been thinking about her, wondering how she's doing, missing her." (My stepmom had told Debbie that Cassie had lived with her sister in foster care.)

"Cassie has a little bit of a loose bladder," Debbie continued. We'd already told her that, too. "She gets a little upset when it happens. Getting her off grains will help with that. But I think it's emotional."

Cassie stretched out on the carpet in front of Debbie. "She keeps asking me about my dogs, 'cause she smells them on me. I keep sending her images of mine — Cassie says she likes hearing laughter." She looked at my dad and stepmom again. "When you guys go out, do you leave the radio on for her?" They shook their heads — they'd never done that. "Hmm," Debbie said. "I wonder if that might have been from another place."

Debbie went on to tell us that Cassie seemed to have problems only with some dogs about her size or larger. "Because every time I send her an image of one of my dogs — I have tiny little furballs — she seems very interested. She seems to have been in a situation where another dog was aggressive toward her."

At that moment, Cassie stood up and stuck her nose in a bowl of bananas and oranges on the coffee table. "She was attacked by a larger dog, a large female dog. Is it mostly large female dogs she has a problem with so far?"

"Yes, so far," my stepmom said.

"Well, we can ask her to change her

behavior," Debbie told us. "Whether she will or not . . . Dogs can be stubborn, just like humans."

The conversation then turned, unexpectedly, to peanut butter. "I have the sensation of peanut butter in my mouth, and that rarely happens," Debbie told us. "Cassie says she loves peanut butter. If you ever need to get something down her, use peanut butter." Cassie jumped up on the couch next to Debbie. "Oh, my, peanut butter is exciting!"

I looked at my dad and stepmom. "Does Cassie like peanut butter?"

"I think most dogs like peanut butter," he said dismissively. "We've given her some a few times, but I can't really remember. There's other things we give her more that really seem to excite her." My dad shifted in his seat and looked at Debbie. "Have we figured out why she's nervous around Benoit yet?" He posed the question as politely as he could, but it was clear he believed we had thus far skirted the real issue.

Debbie probably sensed the urgency in his voice. "What Cassie says is, you know how some people have nervous stomachs? Well, she has a nervous bladder. When she gets overexcited or stressed, the bladder leaks. As she gets older, there will be more strength there."

"We play and roughhouse all the time," my dad said, "but it's only with Benoit that she pees."

"Has this happened with any other person?" Debbie wanted to know.

"The male vet," my stepmom said.

"It's really men that she has a problem with," my dad clarified.

"Cassie's showing me someone, prior to her foster care home, hitting her in the head. It looks like a man to me. I also see her in a crate. Do you know if she was crate-trained?" My stepmom nodded that she was. "She keeps sending me an image of being in a crate, and I hear the reverberations of a metal kind of sound. I think someone might have kicked it, and it scared her."

Debbie suggested that we all walk into the guest bedroom. We'd told her that Cassie wouldn't enter the room if I were in it, though we'd made progress during the two days before her visit. Cassie would stand in the hall outside the doorway, wagging her tail at me; she seemed to be scared of me and excited by me at the same time.

I led everyone — dad, stepmom, and Debbie — into the room. Cassie followed us right in, as if she'd been hanging out with me in there for years.

"What a *good girl!*" my dad said beaming.

"Why do you think she came in this time?" I asked Debbie.

"Cassie says she was unsure of having someone new in here, but that she's feeling better now," she replied.

To her credit, Debbie didn't try to suggest that her presence at the house had caused Cassie to overcome her fear. Instead, she told us, "It's clear to me now that she has gotten close to you, because when we came in she was looking at you for assurance that I was okay to be in here."

Back in the living room, I turned the conversation briefly to my relationship with Casey. "For so much of life, I feel like I've been disappointing him," I told her. "I worry that he's bored a lot, that he's frustrated with me."

When I'd said similar things to other communicators, they'd insisted that while Casey might occasionally feel bored, he loved me to death and wouldn't dream of sharing his life with anyone else. "He cares about you deeply," one had told me. "I'm feeling a lot of love from him. You know when you're really proud of someone and you start to almost well up a bit? That's what I'm feeling from him."

Debbie, though, took a different approach. "You may very well be picking up on his true feelings of frustration," she said, before pivoting to the situation's upside. "It sounds like you're fortunate to have an animal that is good at projecting what he's feeling and getting your attention!"

I supposed that was one way of looking at it.

Though the animal communicators I spoke to didn't always agree on what Casey was thinking or feeling, they were more unified in their analysis of why the universe had paired Casey with me. They insisted that my dog was with me for a reason; he was supposed to teach me valuable life lessons.

Several communicators spoke at length about my dislike of big groups, of my tendency to be a loner. "You can be a non-joiner," one told me. "You'll have one foot in and one foot outside a group. You don't really throw yourself completely in and let yourself be known." Another said, "One of the many reasons Casey is in your life is to get you to go outside and explore the world, to meet people, to take risks."

When I'd lamented to one longtime animal communicator that my dog wasn't meeting my needs (wasn't "offering me unconditional love"), she smiled warmly back at me.

"Well, maybe that's not the reason he was put in your life," she replied.

"What's the reason, then?" I pressed.

"Maybe the reason you're paired with him is so *you* can learn to give *him* unconditional love. Maybe Casey is here to teach *you* how to love."

I'd suspected that was true the second she said it.

8.
In Which I Hang Out with Cesar Millan, Homeless Teenagers, and My Former Middle School English Teacher Turned Dog Masseur

There were times during my journey across America that I felt so deliriously happy — so content, grateful, and blessed — that I considered staying on the road forever.

One of those moments happened on Malibu's Point Dume State Beach, which is tucked away under a promontory at the northern end of Santa Monica Bay. I was walking along the sand with Nic Sheff, a young writer who chronicled his methamphetamine addiction in the book *Tweak*. (Nic's father, David Sheff, wrote his own account of Nic's addiction, titled *Beautiful Boy*.)

It was a glorious day, and Nic had brought along his goofy Bloodhound, Rhett, named after the character in *Gone With the Wind*. Casey and Rhett tumbled around in the sand; Rhett, on his back, pawed at Casey's face. In the distance, Rezzy seemed to be coming alive right before our eyes: she danced along the ocean's edge, her playful personality bursting forth in a joyous mixture of sand, mud, and saltwater.

"If there's anything better than being here right now with our dogs, I'm not sure what it is," Nic said with an easy smile, his curly brown hair falling over his eyes. Those eyes can look vacant and sad in photographs, but on this day they were bright, hopeful. Nic is slender and boyish, and his wardrobe — blue jeans with the cuffs rolled up, thin track jacket, small backpack — gave him the look of a young indie rock star on a walkabout.

"I'm so glad we're here, doing this!" he continued. "Your dogs are awesome!"

"Yours, too!" I said.

We had all the giddiness of starstruck lovers, but we were far from that. Nic isn't gay, and I wasn't interested in him in that way. But our bond was instantaneous and undeniable, perhaps because we have so much in common: We were both raised in the San Francisco Bay Area. We both have divorced parents. We both have writer fathers. We both wrote publicly about our struggles with addictive behavior. And we both have dogs who helped us get better.

In a 2011 article for *The Fix,* an online magazine about addiction and recovery, Nic wrote about the importance of dogs to his sobriety. At the height of Nic's addiction, he wrote, he was homeless "and letting guys blow me for $50 a pop, so I could afford another gram of speed." When his half-brother suggested that the solution to his addiction might be to get a dog, Nic angrily dismissed the advice.

"It just seemed so condescending," Nic wrote. "Like he was totally minimizing my problem."

But several years later, while Nic was drinking "a quart of vodka every day" and "lying to everyone" about being sober after the release of *Tweak,* he came upon an emaciated hound dog running through traffic in Savannah, Georgia, where he was living at the time. Nic brought her to the Humane Society, where she promptly attacked the vet. The vet told Nic the dog would have to be euthanized.

"I could really relate to this crazed, homeless dog, and I felt like she deserved another chance — maybe the same way I still believed I might deserve another chance," Nic wrote in *The Fix.* He didn't let the Humane Society put her down. Instead, he took her home, named her Ramona, "and began the long, slow process of trying to rehab this psycho dog — while, at the same time, I guess, trying to rehab myself."

Before he knew it, Nic had stopped drinking. And though he concedes that therapy and medication helped in that endeavor, he believes his half-brother was right. "I needed to be responsible and accountable for a living creature that literally could not survive if I was off getting fucked up," Nic told me.

Ramona wasn't at the beach with us on the day of my visit; she's still "a handful," Nic said, and occasionally can get aggressive toward people and dogs. "I'd never heard her make a sound until she started

389

growling at me. It sucks when you can barely pet your dog, and she doesn't want to sleep in bed with you. She's even bitten me a few times when she gets anxious."

Though my problems with Casey paled in comparison to the challenge of living with Ramona, Nic was eager to talk about them. Before embarking on my journey, I'd briefly mentioned to him my Casey-related insecurities. "How are things going with you both?" he asked me, sounding genuinely interested.

It was a good question, one I realized I hadn't considered in the two weeks since adopting Rezzy. Though I'd tried not to neglect Casey, Rezzy had commanded practically all of my energy and attention. And, boy, did Rezzy love attention. Even when she was tired (as she was for much of those first two weeks), she preferred to be tired with her head in my lap. She was physical and loving in a way that Casey was only rarely; Rezzy wanted to be as close to me as possible. Nic noticed.

"She's so bonded to you already," he said.

As we sat in the sand watching our dogs, I realized that I hadn't felt any frustration or insecurities around Casey in weeks. "It's almost like rescuing Rezzy made me realize that dogs are different, and that I don't need to expect Casey to be everything," I told him. I was talking out loud, figuring out my thoughts and feelings as they came to me. "And I know I've been paying more attention to Rezzy than Casey, but that's because Rezzy is so new, and she needs me right now. I know Casey is okay."

"Casey seems so easygoing about things," Nic said.

"Exactly. And I love that about him." I paused and let that sink in. "I don't think I've ever realized how much I love that about him. He doesn't even seem to mind the RV anymore. He's happy, he's content. And he doesn't get jealous if I have to pay a lot of attention to Rezzy."

"It seems like Rezzy is the perfect complement to Casey, even down to their colors — black and white," Nic said. We laughed as we watched Rezzy dig a hole in the sand and stuff her nose in it, then run to us through a stiff wind and gently nudge her face in my lap. "She's the most awesome dog. You really lucked out."

"You did, too, with Rhett," I told him, as the dog chased Casey in a circle, Rhett's droopy ears flapping against his head as he bounded through the sand.

"You should have seen Rhett when he was little," Nic said. "He was like this super runt

391

of the litter. Nobody wanted him. So I took him, but he was always sick the first year. I spent so much time looking after him that a few months before my wedding my fiancée was like, 'Why don't you marry the *dog*?' She felt like I was giving him more attention than her. But I was like, 'He's like this sick little puppy, and I have to take care of him.' I had a sick puppy and a psycho rescue. They both needed me."

"And you needed them," I said.

"Yes! There's no doubt in my mind — my dogs keep me sober. They do that by getting me out of myself, by forcing me to think about someone else before me. They make me less self-centered."

Though residential addiction treatment centers haven't included dogs in therapy historically (many use horses instead), some high-end rehabs now allow patients to bring their dogs. Several other rehab centers and sober houses incorporate therapy dogs, or have resident dogs on site.

A friend of mine, Joe Schrank, told me that the sober house he runs — Loft 107 in Brooklyn — would be much less effective without the presence of Churchill, an English Bulldog, and Lucy, a rescue Italian Mastiff.

"When people get out of inpatient treatment and come to us, their lives are usually in shambles," he said. "Everyone they know is angry or hurt, and rightfully so, because addiction is like a tornado through marriages and family systems. Some people are so off their map and shut down, the easy starting point for a relationship is with one of our dogs. The dogs aren't angry with you, they're not divorcing you or threatening to fire you, and they won't hold your past against you." Schrank says the dogs also give recovering addicts a sense that they're "stepping back into the real world rather than being stuck in an institution."

There's a third dog who makes appearances at Loft 107 — a five-year-old black Lab named Mik. Mik's job isn't so much unconditional love as it is unconditional accountability, which might even be more important for addicts in early sobriety. The dog was originally trained for narcotics detection at a small police department in Texas, but he lost his job when the department cut its K9 program. Mik ended up with Joe, who has found the dog to be critical in keeping his seven-thousand-square-foot loft drug-free. Joe likes to say that though his dogs have different jobs, they're all an "addict's best friend."

That may be why so many recovering addicts choose to live with dogs. As Nic imparted to me on the beach, pets that need attention, feeding, and daily walks are a remarkable buffer against relapse. So, too, is the guilt many addicts feel if they drink or use drugs in front of their dogs.

"I'm not sure why, but it's almost worse to let down your dog than it is to let down a friend or family member," a friend of mine who has struggled with drug and sex addiction told me. "The dog is like a mirror — and in his eyes you can see all the sadness and shame you're feeling toward yourself."

Another good friend of mine is convinced that his two dogs know when he's active in his drug addiction. He told me that the younger of the dogs, a Chihuahua/Pit Bull mix rescue, normally loves everybody but would try to chase his drug dealer out of their apartment. The older dog, a tall Chihuahua, would start whimpering when my friend brought out a bag of drugs.

"Then, when I got out a needle, he would make a sound like he only made when he saw that sight — a kind of human cry in a canine throat," he recalled. "It was eerie and heartbreaking. I ended up having to board them whenever I planned to use, because they became like little sheepdogs trying to herd my sober self away from my addict self."

I never got the sense that Casey was trying to herd me away from my addict self in the depth of my struggles, which only added to my belief that he didn't particularly care about me. At my worst in my twenties, I would spend days holed up in my apartment in front of my computer. I used Internet chat rooms and the pursuit of sex as a kind of drug, and nothing — not my family, not my friends, and certainly not my dog — could compete with it.

As my life deteriorated, I took Casey for fewer and shorter walks. I played with him less. I'm ashamed to say that on some days I didn't play with him at all. But unlike my friends and family, Casey didn't have anywhere else to go; he couldn't get away. There is no Al-Anon for dogs. No one told him about the dangers of codependency. He was tied to me, stuck with me. Casey stayed with me back then not because I deserved the company. He stayed with me *because he's a dog.* That's what dogs do, sometimes to their peril. They stay.

Did Casey realize just how depressed I was? There were times when I thought I saw some recognition in him of how small and miserable my life had become. He would look at me with what I was sure were the saddest eyes ever fastened to a dog. He would sigh a lot. But maybe, as my friend had suggested had happened in his case, the dog was serving as a mirror for my own shame.

Here's what I know for sure: Casey helped keep me sane and alive as I lost friendships and a relationship. Though I never tried to take my own life, I certainly thought about it. But the mechanics of dog ownership — walk, feed, repeat — gave me just enough structure and responsibility to feel needed. Besides, I wanted him to see me get better. I wanted Casey to see me do right by him.

The prospect of Cesar Millan meeting my dogs terrified me. "I just know that Casey will be on his worst behavior," I fretted to a friend by phone as I drove to Cesar's Dog Psychology Center in the Santa Clarita hills. I was convinced that Cesar would find Casey boorish and poorly trained. I imagined Casey jumping up on Cesar to say hello, barging through doorways ahead of me (a big no-no, according to Cesar), or, worse yet, humping Cesar's dogs. I worried less about Rezzy, but there was always the possibility that Cesar might discern some failure there on my part, too.

This wouldn't be my first time meeting the Dog Whisperer. A month before embarking on my journey, I'd interviewed Cesar in front of several hundred dog lovers as part of a *New York Times* Arts & Leisure Weekend. Though I'd watched dozens of episodes of his hit show, *Dog Whisperer with Cesar Millan,* I was unprepared for just how funny and charismatic he would be in front of an audience. Joined onstage by his Pit Bull, Junior, Cesar had a comic's timing and an actor's easy physicality. His imitation of the way many Americans walk their dogs — small dog pulling large human — had the crowd doubling over with laughter.

During that interview, Cesar stressed that dogs need three things to be healthy and content: exercise, discipline, and affection, in that order. He told the crowd that dogs in Third World countries tend to be in better shape — physically and emotionally — than dogs here at home, which he believes we treat too often like little people.

"Dogs seem to have everything in America," Cesar said, except an understanding from us humans about what makes them truly content. "We want them to become like us, but it's best for them not to become like us."

According to Cesar, who is famous for

saying that he "rehabilitates dogs and trains people," many canine behavioral problems stem from owners who deal with their pets on an emotional level, ignoring the dogs' pack instincts. Cesar urges owners to be Alphas and to display calm-assertive control over their animals at all times, especially when going for walks.

"Walking in front of your dog allows you to be seen as the pack leader," Cesar wrote on CesarsWay.com. "Conversely, if your dog controls you on the walk, he's the pack leader. You should be the first one out the door and the first one in."

Though Cesar had an adoring audience at the *New York Times* event, some dog trainers and behavioral experts — particularly those who adhere to a positive reinforcement philosophy — disagree with his methods. In a 2006 op-ed in *The New York Times,* writer Mark Derr spoke for many of Cesar's critics when he argued that Cesar is a "one-man wrecking ball directed at 40 years of progress in understanding and shaping dog behavior and in developing nonpunitive, reward-based training programs."

The godfather of punitive dog training was probably William Koehler, author of the 1962 bestseller *The Koehler Method of Dog Training.* The book is packed with nuggets of training wisdom that would be considered wholly unacceptable today. For example, "hold [the dog] suspended until he has neither the strength nor inclination to renew the fight," and "lowered he will probably stagger loop-legged for a few steps, vomit once or twice, and roll over on his side. But do not let it alarm you."

Koehler, who trained dogs for the military and Walt Disney Studios, combined his tough-love approach with a sardonic sense of humor. When describing how to teach a dog to heel, he suggested tying a fifteen-foot leash to the dog and opening the front gate. If the dog tried to run off, Koehler advocated sprinting in the opposite direction. "He'll come with you, if only to be near his head."

Opposition to the pack or dominance theory began to take shape in the early 1980s, under the guidance of trainers like Ian Dunbar and Karen Pryor. "There is nothing that says [punishment] has to be painful or fearful," Dunbar said, "and it doesn't have to be painful or fearful to work, so maybe it shouldn't be." Instead, Dunbar advocates the use of positive reinforcement, where only a dog's above-average behavior is rewarded. With this method, "better responses get better rewards, and the best responses get the best rewards," Dunbar explained. "Then, the learning is ever-learning. The dog is always improving."

In a 2006 piece for *The Woofer Times,* a newspaper for dog lovers in the San Francisco area, Jean Donaldson of the Academy for Dog Trainers called dog training a "divided profession."

"We are not like plumbers, orthodontists or termite exterminators who, if you put six in a room, will pretty much agree on how to do their jobs," she wrote. "Dog training campus are more like Republicans and Democrats, all agreeing that the job needs to be done but wildly differing on how to do it."

In that way, the dog training wars are a lot like the parenting wars. Is firm, old-school discipline the best way to deal with a misbehaving dog — or child? Or is there a better, more modern way? Donaldson told writer Michael Schaffer that Milan's popularity is, in some way, a backlash against modernity and political correctness.

"It's the urge to dominate at least something," she said.

Cesar's Dog Psychology Center sits on forty-three acres surrounded by low mountains and desert palms. He calls it "Disneyland for dogs," and when I got there it was easy to see why. Cesar's pack — there were twenty-two dogs on the day of my visit — lounged around an enclosed gravel dog run, or went for a swim in the bright blue pool at the center of the property. (One of the dogs, a Lab named Holly, is the only animal that's ever sent Cesar to the hospital. She bit his hand after he tried to move her food bowl.)

When the Southern California sun gets too hot, Cesar's dogs can hang out in the compound's air-conditioned kennel. The corn-kernel-yellow building houses a single row of low, gated kennels against one of the indoor walls, each of which has a doggie door leading to an outdoor sleeping space.

Despite the abundance of chain link fencing, which Cesar and his crew use to divide the pack, maintain order, and — most importantly — keep away coyotes, the property is homey and carefully decorated. Each of the outdoor planters is bordered by bright red rocks, which are painted the color of the Grand Canyon. Many contain some sort of ornament: a foot-high Jesus statuette, an imposing ostrich sculpture, or, my favorite, a metal water fountain made to look like a urinating dog. There are less artful, more necessary touches, too; signs remind visitors of the center's three big no-no's: "NO TOUCH, NO TALK, NO EYE CONTACT."

Canines aren't the only species that call the Dog Psychology Center home. At the time of my visit, Cesar owned two large turtles, as well as a horse and a llama. He hoped to add an ostrich to his menagerie. I asked Cesar if his llama had behavioral problems. "In the beginning, he was trying to mount every man he saw," he said. "I corrected the llama, and now I walk the horse, the llama, and the dogs together. It's the most amazing thing."

We spoke in a small white trailer, which serves as his unofficial office, parked at one of the highest points on the property. Cesar wore a blue-collared shirt and a wide-brimmed cowboy hat, under which I could see some salt-and-pepper hair, one of the few traits that betray that he's in his mid-forties. Though Cesar isn't tall, he has crafted an intimidating physical presence. He has a sturdy torso, strong legs, and well-defined arms, and he walks purposefully with his chest puffed out. But there's also refinement and delicateness to him. On the day of my visit, Cesar's teeth were bright white, his mustache and goatee carefully cropped, and his skin tan and glowing.

Though I had many questions to ask Cesar about Casey, I began by talking about my newest dog. "When should I start training Rezzy?" I wanted to know.

"Don't do it yet," he suggested. "Her level of trust has to increase first. Keep it primal. The more chances you give her to walk off-leash in a safe place with you, do that for now."

Cesar then corrected me for having helped Rezzy up the trailer's steps a few minutes earlier. "You carried her the whole way up practically," he said. "In the long run, you're not really helping her if you do that. It would be better if you helped her go half-way, and let her do the rest, because that's how you can help her build self-esteem. At the same time, with a dog that's this shy, if you build too much self-esteem, they can go into a dominant state. It's a fine line." (I couldn't imagine Rezzy going "into a dominant state," but later in my journey Cesar would be proven right.)

I turned the conversation to Casey. Though he's not aggressive, Casey has three behaviors — excessive barking, lunging at dogs when he's on a leash, and humping dogs when he's off a leash — that can get him into trouble (and embarrass me). I was most concerned about his lunging. Though he does it only when he wants to play, it can appear aggressive, especially when combined with a growl that sounds anything but friendly.

"It's the weirdest thing," I said, "because the growl makes no sense when combined with what Casey actually does when he's done lunging. He just wants to say hello and play."

"But other people don't know that your dog is friendly, so a dog lunging at their dog can rightfully seem scary," Cesar told me. "What the other dog sees is that Casey is next to you and then goes in front of you, and in the animal world that's confusing because a follower never removes himself from a follower position. The only reason a follower would go in front is because he senses that no one is in charge of the situation."

"So what can I do to stop the behavior?" I asked.

"What do you do when he's already in that confused leader position?"

"It's a combination of pulling him back and trying to reassure the other person that he's friendly and just wants to play," I said.

"It seems like you're missing a moment where Casey's about to lunge and growl, and you're not giving him the reminder not to do it," Cesar explained. "The human often misses that moment. That's where you can prevent him from going in front. I don't know what his signs are, but a dog might put his head down, or he might lift up a leg, or start pulling on the leash. You have to notice that and snap him out of it. That's how you can use your proximity to him as a tool."

I told Cesar that I wished I could have Casey off-leash all the time, because the lunging and growling rarely happens when he's walking freely next to me. "That means the leash is triggering the wrong reaction," Cesar said. "By doing that on the leash he's telling you, 'I'm naturally a follower, but the leash confuses me and makes me a leader sometimes.' Your job is to learn how to empower yourself to keep him in a follower state on the leash." (Cesar often laments that many humans don't know how to walk a dog. "I have clients who are Harvard graduates," he told me at the *New York Times* event, "but they can't walk a Chihuahua.")

A few minutes later, Cesar took Casey, Rezzy, and me to meet his pack. "Now, we get to see the truth about your dogs!" Cesar said with a smile.

"They haven't had a walk today yet, so

they might be a little rambunctious," I told Cesar, trying to prepare him — and probably me — for the worst.

Cesar opened the fence to where his dogs were hanging out. Casey sauntered right in, while Rezzy stood back and watched.

"She's a little scared," I said.

He shook his head. "No, she's not scared. Actually, that's respect. She's more in the normal state. Casey can get in trouble by being so cocky."

Casey promptly started humping Holly, the Lab who once bit Cesar. "He's going to try to establish himself in this new pack," Cesar said. Fortunately, Holly is only aggressive toward humans who dare to touch her food. "If he tries to reinforce dominance with one of these guys" — he pointed to two of his bigger dogs, including a Mastiff/Lab mix named Joe — "it's going to be a big, big problem."

On cue, Casey started humping Joe. Cesar grabbed Casey's collar and pulled him off, but Casey wouldn't take no for an answer. When Casey tried again, Cesar reached out reflexively and pushed Casey away with his fingertips — so quickly it looked like Cesar had zapped him with a tiny electric shock. (He hadn't.) Casey barked at Cesar — a *why are you ruining my fun?* bark — but then backed off. Fortunately for Casey, Joe happened to be in what Cesar called a "calm-submissive state." The dog also had diarrhea and seemed more intent on finding a place to poop.

A few seconds later, Casey tried mounting another of Cesar's dogs. Cesar pulled Casey off again and then cornered him for a few seconds until he calmed down. "He doesn't know how to go into a total calm-submissive state," Cesar said. "It's like you put a pause on a tape, and then he goes right back."

"When he does this in a dog park, I pull him off and tell him, 'No,' " I said.

"But that's just a temporary fix. You have to do the follow-through, where the brain becomes submissive, and then he associates the touch of the collar, or the 'No,' with that state of mind."

I've never been very good with humping follow-through. Part of the problem was that in Boston's Franklin Park, where I walked Casey for years, many of the dog owners didn't mind my dog's clumsy attempts at dominance. They would laugh them off and act as though I was overreacting if I pulled Casey off and scolded him. I got so used to the lax attitude that I wasn't prepared when someone — usually new to owning a dog — would shriek in horror at the sight of their pet being mounted by mine.

A surprising number of dog owners assume that a humping dog wants sex with their dog, when in fact a humping dog is usually trying to establish dominance. At an RV park in Florida, one man — who was there with his very large male Mastiff — pulled Casey off his dog, shot me a dirty look, and said, "That's illegal in this state, son." Then there's the Tennessee man I read about who relinquished his male Bulldog mix to a shelter because he found the dog "hunched over" another male dog — and assumed his pet was "gay."

But sometimes people have a good reason for not wanting their pets humped. Some dogs react aggressively to being mounted; others have bad hips. Casey, though, doesn't seem to care whether I scold him or not. He loves to hump. And he was in rare form at Cesar's.

"He's not that dominant a dog — that's what's so weird about his humping," I told Cesar as we walked back toward the center's main administrative building, where he had a meeting.

"But because everyone else here is more submissive than him, it's just natural," Cesar said. "I just keep my dogs as submissive as possible, so that your dog will come out of here alive."

I appreciated that.

Before saying good-bye, I asked Cesar if he had any advice for the remainder of my trip.

"Have fun!" he said. "If I wasn't so busy, I'd come with you."

The best way to travel from Los Angeles to San Francisco on Route 1 is probably in the passenger seat of a convertible.

The famed coastal road, which twists and turns as it connects Northern and Southern California, boasts brisk ocean air and some of the most idyllic scenery imaginable: forests of skyscraping redwoods, hidden-away beaches broken by stone cliffs, and gleaming sapphire water.

Route 1 also boasts what *National Geographic* rightly calls "hair-raising drop-offs" into the water, as well as the occasional landslide and falling rock. Near Big Sur, especially, the mountains don't just appear to sag and drop off into the Pacific — sometimes, they actually do.

All of that is to say I was grateful to be driving *north,* meaning there was a lane of traffic between the chalet and an obituary that read, "Benoit Denizet-Lewis, a writer traveling around the country with his two dogs, died Thursday when his motorhome careened off a cliff into the Pacific Ocean." Though I'd made the drive before, I'd somehow forgotten how easy it was to get distracted by the views around each sandstone cliff.

Casey and Rezzy slept for most of the drive on that beautiful mid-April afternoon. During the previous week, they had devised something of a system. If they both curled up just so, they could fit — barely — in the small space between the passenger and driver seats. Other times, they wouldn't get it quite right. If Casey got there first and didn't leave enough room for Rezzy, she would simply lay half of her body on top of his, causing him to sigh loudly and eventually stand up in frustration. If Rezzy snagged the space first, she'd spread out, leaving Casey to stare at me as if I needed to correct the injustice.

I wasn't in the mood to play doggie musical chairs with Casey and Rezzy as I navigated the RV through the more treacherous turns on Route 1. In Big Sur, I pulled over at a turnout near the Bixby Bridge (one of the highest single-span concrete bridges in the world) and buckled Rezzy into a doggie seatbelt harness in the passenger seat. Sadly, she didn't love it there. Like Casey, Rezzy coveted the small, secure floor space between the seats.

I called Marc from the turnout. He was at the Fort Lauderdale airport, waiting for his flight to San Francisco to spend a few days with me and the dogs. I was eager to see him again, show him where I grew up, and introduce him to Rezzy. I'd chosen to take the high road and not remind him that he'd hinted that I should leave her in Kaibeto.

In the Bay Area, I drove north of San Francisco to a KOA in Petaluma. We were further from the city than I would have liked, but the upside was that we were closer to Napa Valley's wine country.

Marc had never been to San Francisco before, so for a few days I got to play tour guide. We drove around in a rental car, the dogs in the backseat. Casey, having finally gotten used to the RV, now seemed uncomfortable in a regular car.

In Sonoma, we got a private tour of the Saint Benoit Creamery, which gave us bottles of yogurt for the road. In San Francisco, we played with the dogs in Golden Gate Park, walked them through the neighborhoods where I grew up, and visited my former middle school English teacher, Roger, who now works part-time as a "dog masseur."

As Roger massaged Casey, he told me that he came upon this line of work by accident. "I made the business cards as a kind of joke," he said, "but there are so many crazy dog people in San Francisco that people took me seriously. And I do like petting dogs, so it works out nicely." I asked Roger, a lifelong dog lover, why he didn't have one of his own. "I've always chosen to live in buildings that don't allow dogs," he explained. "I know myself, and I know that if I had one, I wouldn't bother hanging out with humans anymore. I would end up in a monogamous, nonsexual relationship with a dog."

At the end of our first day in the city, we drove to the SOMA neighborhood and met pet photographer Amanda Jones, who'd photographed Casey early in my journey and who happened to be in San Francisco for a weekend shoot. I was looking forward to having her take pictures of Rezzy, but to everyone's surprise my newest dog despised the attention. I wasn't sure what terrified her more — the bright lights or the clicks of Amanda's camera — but she wouldn't stay still and kept trying to slink off the set. I managed to bear-hug Rezzy into submission for a few pictures with me and Casey, but soon I felt like a parent trying to force his tomboy daughter into a child beauty pageant.

Two days later, we all returned to San Francisco to spend time with one of the city's greatest treasures — Armistead Maupin, author of the *Tales of the City* series. A dog lover, Armistead had heard about my journey and was following my progress on Facebook.

On a cool, cloudy, windy afternoon — the kind San Francisco should bottle and distribute under the name *Sweater Weather* — Marc and I drove the dogs to Crissy Field beach, not far from the Golden Gate Bridge, to meet Armistead and his husband, Chris. They were there with their hyperactive Labradoodle, Philo, named after Philo T. Farnsworth, the inventor of the television. (Philo appears in Armistead's last two novels — *Mary Ann in Autumn* and *The Days of Anna Madrigal* — as a Labradoodle named Roman.)

After playing with the dogs at the beach, Marc and I followed Armistead and Chris to their home across town for a leisurely dinner. There, our conversation soon turned to gay men and their dogs. I told everyone that two of my favorite books — *My Dog*

Tulip, by J. R. Ackerley, and *Dog Years,* by Mark Doty — happen to have been written by gay men. (Ackerley's dog was actually named "Queenie," but when the book was published in 1956, there was apparently little appetite for a book titled *My Dog Queenie.*)

I told Armistead, Chris, and Marc that I'd just finished reading *Paws and Reflect,* a collection of essays that "celebrates the special and powerful bond" between gay men and canines. For many of the *Paws and Reflect* contributors, a pet dog provided a much-needed respite from the secrecy, shame, and arduous self-analysis of growing up gay in a straight world. Alone with their dogs, the authors could let their guard down and be themselves. They didn't need to keep secrets from their dogs, nor did they have to worry that their dogs would judge them — or, worse yet, reject them.

As adults, many gay men — whether single or partnered — seek out the company of dogs. For the many childless gay couples in America, canines can take on the role of surrogate children. And Donald Hardy spoke for many single, dog-loving gay men the world over (and probably single, heterosexual women the world over) when he wrote, "(Dogs) are a lot like boyfriends, actually, only nicer. And usually they stick around longer."

That's not to say that every gay man makes a quality dog companion. In *Woof!: A Gay Man's Guide to Dogs,* Andrew DePrisco quips that some gay men "can't commit to a color, much less a living creature." And Armistead told me that his longtime friend, the late writer Christopher Isherwood, "maintained that a pet would deflect the love passing between partners." Armistead told me it was one of the few times he ever disagreed with Isherwood.

Armistead has had four dogs in his adult life, including a Poodle named Willy about whom he penned a charming essay titled "Kiss Patrol."

"Some dogs, I'm told, like to stick around when their owners are making love," Armistead wrote. "They'll sit stone still and watch the proceedings with deadpan intensity, as if collecting evidence for some evil congressional subcommittee. Not Willie. As soon as human passion rears its ugly head — and he has an uncanny eye for the precise moment — he flings himself off the bed and skulks away to another room. This is jealousy, I suppose, mingled with mortification, though I'd like to believe there's an element of courtesy involved as well."

Rezzy had exhibited no such restraint the previous night when Marc and I had enjoyed some private time in the Chalet. She'd watched the proceedings from the foot of the bed, seemingly mesmerized. She eventually hopped up on the mattress, nestled herself between Marc and me, and tried to plant kisses on both of our faces.

Though it was one of the funniest things I'd ever seen, I didn't want to encourage the behavior. "Manners!" I protested, elbowing Rezzy toward the edge of the bed. I couldn't help laughing, though, and she didn't appear to take my admonishment seriously. (Her tail thumped up and down against the bed.) Seconds later she was back at it.

"Rezzy, no!" I said, more forcefully this time. I tried to knee her out of the way, but she made herself heavy and wouldn't budge.

"Don't make me put you in your crate," I threatened, though Rezzy was unaffected by my tough talk. In fact, she almost seemed to be smiling as she stared at Marc.

"I guess this is when you realize you rescued a weird dog?" Marc said, amused by Rezzy's apparent crush on him.

"I know! Casey usually just sleeps."

"And I do love that about Casey."

"I think Rezzy just senses that closeness

and intimacy are happening, and she wants to be a part of that," I said.

"Or maybe she's into gay guys," Marc joked.

The anthropomorphic possibilities were endless, and Rezzy took advantage of our momentary distraction to inch toward us on the bed.

"She's a sly one," Marc said.

"A sly one who's getting a *timeout*," I announced as I jumped out of bed, picked Rezzy up in my arms, and marched her to the front of the RV. On my way back to the bed, I pulled the privacy curtain separating the sleeping area from the rest of the motorhome.

"Now, where were we?" I said, with as much sexiness as I could muster.

After a leisurely week in the Bay Area, I had to get back on the road. I didn't want to see Marc go, but we'd made a plan to see each other a few weeks later in Chicago, one of my last stops on the trip.

First, though, I had dogs and humans to see in Oregon and Washington. On the last day in April — a clear, sunny afternoon in the Bay Area — I packed up the RV and drove through Mendocino, then along the Avenue of the Giants in Humboldt Redwoods State Park. I stopped to check out one of the scenic highway's most famous sights — a 950-year-old, 250-foot-tall redwood that has survived logging, a flood, and a lightning strike, earning it the name the Immortal Tree. The park also boasts redwoods so large that visitors can drive cars through their trunks. None, though, seemed wide enough for the Chalet.

I spent that night at the Shoreline RV Park in Eureka, a coastal town favored by dog lovers, hippies, and dog-loving hippies. (At Shoreline, a sign on the front desk read, "Pets welcome. Humans must be on leash.") The next morning, I took Casey and Rezzy for a walk through the RV park. Rezzy had learned to fetch tennis balls by watching Casey, and even with her slight limp, she was still faster than he was. To help Casey out, I would throw one ball for Rezzy, and while she was distracted chasing it, I would toss another one in the opposite direction for him.

Until then, I'd mostly let Rezzy off the leash only in enclosed areas — dog parks, parking lots, Little League baseball fields. (I'd lost count of the number of baseball diamonds I'd passed on my journey. And miniature-golf courses.) But I took a chance that morning, tossing balls for the dogs in an open grassy portion of the RV park. At one point, I noticed Rezzy peering through some bushes that separated us from an adjacent field of knee-high brown grass. Before I could say "Don't you dare!" she made a run for it, dashing through the brush.

She ran so fast! As I chased after her through the field, angry at her and in awe of her at the same time, I tripped and face-planted into a shallow pool of mud. Rezzy probably thought this was a new fun game — *slow human struggles to chase fast dog through muddy field!* — but when she reached a waterway she had nowhere to go. So she sat down and looked back at me as I stumbled toward her, cursing her name.

Rezzy wagged her tail as I scolded her, which only made me angrier. For all my complaints about Casey, he had never run away from me. And, if he had, he'd certainly know to look guilty when I caught up to him. Speaking of Casey, where was he? I looked back across the field and saw him in the distance, staring out toward us. *What a great dog he is,* I thought to myself. In that moment he seemed like the most well-behaved animal in the world.

I hosed off Rezzy, took a shower, and got back on the road. Well after I'd crossed the state line into Oregon, dense redwood walls guarded the slow northeast curve of the aptly named Redwood Highway. By the time I stopped in Trail, an unincorporated community in the southwest part of the state, it seemed I'd seen the full spectrum of green — miles of grass, shrubs, and evergreens all came in different shades of the color.

I spent the late afternoon and evening at Trail's Bear Mountain RV Park, across the road from the Rogue River, which flows westward from the Cascade Mountains to the Pacific Ocean. As I walked the dogs around the sleepy property, where dozens of motorhomes seemed to be in permanent residence, I noticed a neighboring yard with dogs, chickens, sheep, and llamas. I walked over and introduced myself to the proprietors, a retired military man named Merle and his younger wife, Robin, who wore a T-shirt that read, "I'll try to be nice, if you try to be smart."

Over drinks in their living room, I asked them how they'd ended up with so many animals in one yard.

"Well, that's a funny story," Merle said. "Originally we were just trying to get the grass cut, so we got the sheep. But then coyotes went after them, so I got llamas to protect the sheep. But then cougars started coming around, and llamas don't protect against cougars, so then I got a couple guard dogs. There's also a bear that comes around, but all he ever does is eat my grapes."

Merle added that all his animals were rescues.

"Even the llamas?" I asked.

"Even the llamas," he confirmed.

"What about the chickens?"

"Okay, so not the chickens," he conceded.

Merle and Robin had five dogs, including a fourteen-year-old black Lab named Pal they'd found tied to a mailbox after Hurricane Katrina when they lived in New Orleans. Pal spends his senior years in Oregon digging for gophers, and a few days before my arrival he'd gotten a rude surprise when he pulled his nose out from a gopher hole.

"A gopher was hanging on to his face!" Merle said with an excited clap of his hands. "Pal's been pissed ever since."

The dogs — who range in age from one to fourteen — play with all the animals, including the chickens. "I try to keep the dogs from the chickens, because the chickens don't like that so much," Merle told me, adding that he has one "Alpha chicken" who thinks he's a dog and is known to get away and scamper around the RV park. "He's a mean, mean chicken! He's always bossing around the other chickens and trying to get into the dog food."

"He sounds like a nightmare," I said.

Merle laughed. "A funny nightmare, though. I'm not sure what's going on his brain, but that chicken provides us endless entertainment."

The next morning I drove through Crater Lake National Park on my way to Bend, the largest city in central Oregon.

I had lunch in Bend with Kelly Ausland of freekibble.com, which has donated more than ten million meals to animal shelters and food banks across the country. We were joined by Lynne Ouchida, who'd brought along her three-legged disc dog, Maty, the first tri-pawed dog to qualify and compete in the Hyperflite Skyhoundz World Canine Disc Championship. An Australian Shepherd mix, Maty had been abandoned in a hotel when she was only three weeks old. At eight weeks, an infection ravaged her body and caused her to lose her rear left leg.

After lunch, Lynne took us to a park so we could watch Maty in action. While Rezzy and Casey chased tennis balls, Maty leaped high into the air after a disc until a park employee showed up to ruin the fun. "I'm going to need you all to leash your dogs," he said, seeming to enjoy being the bearer of bad news. It was, remarkably, the first such admonishment of my journey.

Lynne had alerted a local television station about my cross-country dog adventure, and to my surprise a camera crew showed up to interview me. I'd go on to receive more media attention two days later in Portland, when another local station came calling. I'm not sure what that says about Oregon; either nothing much happens there, or, more likely, the state really loves dogs.

Just about any Oregon city could make a "most dog-friendly place in America" list — Portland, in fact, routinely does. That city's Pearl District, in particular, might just be the most dog-crazed neighborhood in the world, prompting backlash from those who prefer shopping for groceries without navigating their carts around four-legged animals.

Vance Bybee, head of the food safety division of the Oregon Department of Agriculture, told *The New York Times* in 2009 that customer complaints about dogs in supermarkets were increasing. "Usually they'll hold off . . . until they've seen a dog urinate in the grocery store or jump up and try to swipe a pack of meat," Bybee said. "Or they've seen dogs pooping in the aisle, that sort of thing."

The complaints prompted Oregon officials to pass out posters and flyers to thousands of retail stores that sell food, reminding consumers that only service dogs were allowed inside. Still, the exception is easily abused. In recent years, a number of dog lovers without disabilities have taken to fraudulently passing off their pets as service animals, with some going so far as to purchase fake service vests.

As Tara Palmeri discovered during an undercover investigation for the *New York Post* (she outfitted her mom's dog with a fake service jacket), the "craziest, most badly behaved sons of bitches can run wild in the most elegant eateries in town — as long as they're masquerading as service dogs."

I'd spent time with several people with service animals during my cross-country journey, including perhaps the most dog-obsessed woman I'd ever met — Beth Joy Knutsen. I had first become aware of Beth, a former *Seinfeld* Elaine impersonator and contestant on the short-lived reality show

Greatest American Dog, at a 2009 pet industry event in Las Vegas. She was there hawking her dog clothing line and doting over her little ten-year-old rescue dog, Bella Starlet, who also goes by the names "Bella Bear," "Cuddle Muffin," "Puppy Monkey," "Care Bear," and "Baby Girl."

Beth considers herself a "stage mom/manager" for Bella, a fifteen-pound white mutt who has starred in commercials (including a Dr. Pepper Super Bowl ad), headlined the all-dog float in New York's Village Halloween Parade, and inspired a Soulmates: Mom and Me! jewelry collection for dogs and their owners. On Bella's web page, Beth makes clear that her celebrity pup appreciates the finer things in life: "She enjoys prancing down red carpets where she impresses the paparazzi with her jump-turns."

One doesn't need an advanced degree in psychology to wonder if Beth, whose acting career never took off, is trying to become famous vicariously through her dog. But Beth relies on Bella for more than ego gratification. Beth told me that she suffers from gallstones, and that Bella, a registered service dog, will bark if Beth's body temperature goes up — a sign that an attack could be imminent.

"Bella is my doctor, my baby, my best friend, and my big celebrity all in one adorable little package!" Beth told me the first time we met. A few seconds later she added, "Isn't this the most pawwwwwwwsome day?"

In addition to Bella, Beth has two other dogs, including a therapy animal named Riff. Beth often takes Riff and Bella to classrooms at a local elementary school as part of the school's therapy dog program. She shows the kids episodes of Bella's cartoon, *Pup American Idol,* which teaches life lessons ranging from treating animals kindly to avoiding peer pressure and being yourself.

Therapy dogs like Riff are used in practically every therapeutic setting one can imagine in this country, including nursing homes, hospitals, prisons, and schools. Specially trained dogs help relax veterans with PTSD, comfort kids testifying in court, teach prisoners responsibility and empathy, distract Alzheimer's patients during periods of agitation, and help autistic children develop social and motor skills. (In Phoenix, I'd shadowed a Golden Retriever therapy dog named Tucker as he played with kids at an Early Head Start Program in a low-income neighborhood. Tucker was affiliated with Gabriel's Angels, an organization that brings therapy dogs to some thirteen thousand kids and teenagers each year in Arizona.)

The first therapy dog came upon her job by accident. At the height of World War II, an American soldier found a shivering, four-pound Yorkshire Terrier in a foxhole during a battle in a New Guinea jungle. He rescued the dog and sold her to Corporal William Wynne, who named her Smoky. When Wynne was hospitalized with dengue fever in 1944, his physician, Dr. Charles Mayo — namesake of the famed Mayo Clinic — allowed Smoky to visit him in the ward.

"I was thrilled, naturally," Wynne, who is now in his nineties, told me. "I said to the nurses, 'I don't feel sick anymore!' — and it was true." Suspecting that Smoky might help other recovering soldiers, Wynne's nurses asked permission to bring the Yorkie with them on rounds. They were on to something: according to Wynne, wounded veterans would argue across hospital beds about who was going to get to spend time with the lapdog.

After my television interview in Bend, I drove the RV across town to Sage Veterinary Alternatives, where vets treat animals with acupuncture and Chinese herbs in addition to traditional Western techniques. (It wasn't my first foray into alternative dog treatments. In Palm Bay, Florida, Casey had received an "energy treatment" from Reiki practioner Marie Conrad. Casey seemed to enjoy the attention, though only he knows if the session helped heal any physical or emotional doggie wounds.)

I'd hoped to learn more about veterinary holistic medicine, which has been on the rise in the United States since the early 1990s, but I was mostly concerned about Rezzy. Her bum leg seemed to be getting worse. More immediately, I hoped for a solution to a new problem that had begun only days earlier: Rezzy wouldn't stop itching her stomach and sides. I'd convinced myself that I'd given her whatever had ailed me.

Vet Steven Blauvelt invited me into his bright, homey office, where a gigantic dog bed took up half the floor. He then quickly put my transmission fears to rest. Rezzy's problem was a mild staph infection, probably caused by stress.

"She was recently pregnant and then she got rescued and put in this new environment," he said after taking a few minutes to examine her stomach. "Without getting too esoteric, her body's going through a kind of detox right now."

Next, he turned his attention to Rezzy's leg. He took his time, squeezing and prodding her knee. As he did, he told me that one of the ways he differentiates himself from traditional vets is by spending longer with his patients. "A typical vet's appointment will be fifteen minutes, maybe thirty," he said. "My appointments last an hour. I try to take the time to really understand the specifics of each dog's situation and to understand the dog's lifestyle and diet, as opposed to saying, 'Oh, just another dog with arthritis.'"

Steven was friendly and self-effacing, and he apologized several times during my visit for "boring me" with the details of his holistic practice. "I know I'm not nearly as exciting as some people you've met on your journey — I'm definitely no *Cesar Millan*," he said.

"Nonsense," I told him. "If I lived in Bend, you would be my vet." I couldn't get over how welcoming and un-vet-like his office was. "It's the kind of place where dogs — hell, even humans — would want to hang out."

"That's the point," he said. "A vet's office doesn't need to be an unwelcoming place."

(Since my visit, Steven opened up his own practice, Four Paws Wellness Center.)

After a few minutes, he gave me his verdict about Rezzy's leg. "I'm pretty sure she has a partial ACL tear," he said. "This is kind of the bane of a vet's existence. In the case of old dogs that come in with this kind of injury, we'll do palliative care with acupuncture, laser, whatever we can besides surgery. But if I see this in younger or even middle-aged dogs, I'll always try to refer them somewhere for surgery."

"How much will the surgery cost me?" I asked.

"You'd better start selling some books," he said.

On my way out, he handed me a bag with several medications, including an antibiotic, a bottle of gentle shampoo, and fish oil — all to help Rezzy's skin problems. To combat any inflammation in Rezzy's knee, he gave me an herbal anti-inflammatory.

"Is this like ibuprofen?" I asked.

"Think of it as *herbal* ibuprofen," he said.

For my final stop on the West Coast, I was fortunate to be offered a room at Seattle's iconic and dog-friendly Sorrento Hotel.

It was nice to get out of the Chalet for a few days. I hadn't bothered to clean it since

Marc left, and it was looking eerily reminiscent of the sleeping quarters in my college fraternity house. But the posh Sorrento, which was built in 1909 in Italian Renaissance style, felt like an odd backdrop for what I'd come to the area to explore: the lives of homeless people and their dogs.

Dogs, of course, are far less superficial than we are. They don't care what we look like, how much money we make, or even whether anyone else seems to like us. According to the *Odyssey,* when Odysseus arrived home after twenty years disguised as a beggar, he was first recognized by his old dog, Argos. As Aldous Huxley once said, "To his dog, every man is Napoleon; hence, the constant popularity of dogs."

This is especially good news for the homeless, who rely on their dogs for companionship, protection, and human interaction (having a dog stimulates conversation with the public). What does the dog get in return? Virtually all his time spent outside with a human who never leaves to go to the office.

Cesar Millan told me that the dogs of homeless people tend to be content and well socialized. Mark Wells, an investigator for the Oregon Humane Society, agrees. "These are some of the most loyal, best socialized, friendliest dogs I see," he told *Willamette Week*.

But don't tell that to the many concerned dog owners who call animal control agencies demanding that the dog of a homeless person be "saved" from its circumstances. "It just goes to show you how much we don't understand what a dog needs," Cesar Millan told me. "I'd much rather *save* a spoiled dog in Beverly Hills who practically never touches the ground."

The unique relationship between homeless people and their dogs isn't without its challenges, though. Few cities offer free vet care for homeless pets, and if a dog gets separated from its owner, there often isn't a way to reunite them. More importantly, having a dog can actually prolong homelessness. Because most homeless shelters don't accept pets, many homeless people choose to stay on the streets rather than give up their dogs. The same goes for more permanent housing. In one study, more than 90 percent of homeless people with dogs said they would turn down an apartment where pets weren't allowed.

The same study found that homeless pet owners were not markedly more depressed than dog owners with housing. That came as a surprise to the study's authors. "Possibly," they speculated, "animals buffered the stress of homelessness and increased contentment in the homeless state."

During my second day in Seattle, I drove the RV to a park in Kent, an outlying suburb south of the city. There, on a patch of grass near the city's library, I spent the afternoon with a homeless teenage couple — Dakota and Blueberry, both nineteen — and their eight-month-old black Lab/Pit Bull mixes, Marley and Calypso. I'd been introduced to them by Carey Fuller, a homeless woman who lives out of a car and writes about homeless issues. She's a guardian and mother figure to many of the homeless youth in Kent.

On the day of my visit, Dakota and Blueberry (her real name is Amanda, but everyone calls her Blueberry) had just come from their secluded homeless tent camp near a golf course in an unincorporated part of the city. A local church was delivering lunches that day to the homeless, and as we spoke dozen of teens and adults — some drunk or high — lounged shirtless in the sun near train tracks, paper plates strewn around them.

Blueberry, who looked almost angelic with her pale skin and striking blue eyes, told me she'd been homeless since she was fourteen. "It was better than being around my mom, who's a tweaker," she said, meaning a meth addict.

Blueberry lived on the streets of Seattle until the year before my visit, when she came to Kent and fell in love with Dakota. Though Dakota didn't make the best impression during their first extended conversation (he was on a "robo-trip," a high induced by purposely ingesting too much cough syrup), Blueberry said she could see through his drug-induced silliness.

"I saw a lot more in him than what he was showing in that moment," she told me.

Dakota, who is tall and lanky and wore an oversized T-shirt and baggy jeans, has been homeless on and off since he was eight. (His parents are also drug addicts, Fuller told me.) Dakota used to get high all the time, too. His drugs of choice were meth, ecstasy, and cocaine, and he told me that a doctor once expressed surprise that Dakota was still "alive and walking around." Dakota insisted he only smokes weed these days and is mostly concerned with spending time with Blueberry, looking after their dogs, and securing his GED.

Both Blueberry and Dakota were outgoing and friendly with me, and they seemed excited by the prospect of someone writing about their dogs. "Aren't they beautiful?" Blueberry said with a childlike wonder as the animals rolled around in the bright green grass, offering up their bellies. I couldn't get over how young Blueberry looked — she could have passed for fifteen or sixteen.

"Where did you get the dogs?" I asked her, while Dakota lit a cigarette.

"We were hanging out with my mom and a bunch of tweakers, and some random Mexican just dropped the dogs off," Blueberry said. "So we took 'em."

The dogs are brother and sister; Marley is especially attached to Dakota, while Calypso rarely leaves Blueberry's side. "We're trying to work on her separation anxiety," Blueberry told me with a smile. "She'll start to whine if I go too far. But she's getting a lot better. I know that people can't imagine that dogs living on the street are spoiled, but these two . . ."

"They're *so* spoiled!" Dakota interjected.

"They get so much attention and love, it's ridiculous," she continued. "Everyone wants to pet them."

"And they always eat before we do," Dakota was sure to say. "I can go a few days without eating, as long as I have water. Any money we get from spanging" — begging for change on the street — "goes to the dogs first." (Many of the local churches that deliver food to the homeless also hand out dog food.)

Blueberry and Dakota planned to make a trip to Seattle, to a church that offers free shots and dog licenses. Marley and Calypso also needed to be fixed. "The last thing we need are a bunch of little puppies running around," Dakota said.

It was warm during my visit, so to keep the dogs hydrated Blueberry and Dakota kept filling a plastic Starbucks cup with water from a jug. When the couple had to run to the library for a few minutes to check their email (both were hoping to hear from prospective employers), two homeless young people volunteered to watch the dogs. There's never a shortage of dog-sitters for Marley and Calypso, who are friendly and only "bark at tweakers," Blueberry told me.

When they returned, Dakota and Blueberry explained that their dogs offer them a priceless commodity on the streets: protection. "The other night some people kept getting close to our tent, but the dogs scared them off," Dakota told me. "No one messes with us when we have dogs."

(I was reminded of Lars Eighner's mem-

436

oir, *Travels with Lizbeth,* in which he credits his dog with helping to keep him safe during his three years of homelessness: "Anyone who has had to sleep by the side of the road in some wild place may appreciate that an extra pair of keen ears, a good nose, and sharp teeth on a loud, ferocious ally of unquestionable loyalty have a certain value that transcends mere sentiment.")

Some of the homeless youth in the Kent area are as young as twelve or thirteen, and Blueberry and Dakota told me they could all use a dog for security. "There are crazies and pedophiles out here that a dog can protect you from," Dakota said.

"But most kids that age aren't responsible enough to take care of a dog," Blueberry told me. "I know I wasn't when I was their age."

It just so happened that on that day, seemingly everyone in this homeless community was on the lookout for a man who they said had raped a young homeless girl in the area.

"A few people have seen him hanging around the last few days, trying to blend in," Blueberry said. "Everyone's looking for him."

Every once in a while, Dakota would wander off in a pack of other homeless youth in search of the man.

■ ■ ■ ■

Later that afternoon, Dakota and Blueberry pointed out four homeless kids shuffling along the pavement next to the library. Each wore baggy clothes and backpacks; the youngest of the four couldn't have been older than twelve.

"Three of those kids are related — they just started coming around," Dakota explained. "They kind of keep to themselves, which is normal for being new out here."

Earlier, Fuller had told me about a twelve-year-old who'd arrived the month prior after running away from a drug rehab facility. "We were all trying to get him back home," she said. "If it's not a violent situation, everyone will try to get those kids back with their families."

For Dakota and Blueberry, going home isn't an option — it's "better for our mental health" to be out here, Blueberry said. A homeless shelter isn't a possibility, either. "They don't take dogs," Dakota explained, "and I got beat up and robbed at the one I did go to."

That's not to say that Dakota and Blueberry like living in a tent. The stress can be unbearable. "For a while," Blueberry told me, "Dakota and I were having a lot of problems with our relationship, in addition to all the other things we have to worry about out here. But the pups" — she looked down at Marley and Calypso, who relaxed at their feet in the grass — "helped us work it out. I'd take Calypso and go in one direction, and he'd storm off with Marley. They helped us communicate, because we would each talk to our dogs, and that helped us then come back and talk to each other."

Dakota nodded in agreement. "The dogs make us realize that we were just being dumb, that we loved each other and could always work it out."

Blueberry told me that she's been diagnosed with several psychological conditions, including depression, insomnia, and ADHD. "The whole nine yards," she said. "When I get really low, I don't like who I am or what I do."

She recalled a recent incident when she visited her mother and felt suicidal. "Every time I was about to do something harmful to myself," she told me, "Calypso would pop up out of nowhere and sit on my lap and kiss me and make me love her. It was the first time I was ever able to get instant relief like that. She saved my life."

Toward the end of my time in Kent, I

overheard Dakota and Blueberry talking about saving up for a hotel room for the night. They wanted a soft bed and a shower. They hadn't asked me for money all day, which was a change from when I'd spent time with homeless youth in San Francisco for an article a decade earlier. Those kids — many of whom were drug addicts — had never stopped asking me for cash, though I was hesitant to give them any when I knew what they'd spend it on.

Even though Dakota and Blueberry didn't ask me for a dime, I handed them $80 for a hotel room. They were good kids who were out here because their parents had failed them, because we as a country had failed them. My gift wouldn't do much to alter the course of their lives, but it would give them one night of peace and quiet.

"Thank you so much!" they said over and over again as I climbed back in the RV.

"Where will you stay?" I asked them.

"There's a dog-friendly hotel nearby," Blueberry assured me.

"I'll be in a dog-friendly hotel, too," I told them.

Dakota looked confused and pointed toward the Chalet. "But don't you sleep in there?"

"Normally, yes," I said. "But I was offered

a room for the night."

"Anytime you get sick of your RV, you can bring it around here," Dakota suggested. "We'll pack a bunch of homeless kids in there. It'll be a party."

I laughed. Blueberry smiled. Then I wished them luck, hugged them goodbye, and watched as they shuffled down the street at sunset, their dogs prancing along at their sides.

9.
IN WHICH WE HIGHTAIL IT HOME, WITH SOME HEARTBREAK ALONG THE WAY

In Idaho, I slept in the belly of a Beagle.

The Cottonwood Dog Bark Park Inn Bed & Breakfast, in the north-central part of the state, is a thirty-foot-tall wooden dog with fourteen-foot carpets for ears. Painted black, brown, and white, the Beagle goes by the name Sweet Willy and can be seen from U.S. 95, Idaho's primary north–south highway. The inn harks back to a time when travelers would often fill their gas tanks or stay overnight in novelty structures: teakettles, seashells, airplanes, elephants.

I can't remember where I'd first read about Sweet Willy, which its creators — the husband-and-wife team of Dennis Sullivan and Frances Conklin — built and lovingly call "a noble and absurd undertaking." But I couldn't pass up the chance to spend a night in the doghouse.

Frances wasn't there when I rolled into town, so I found Dennis alone in his woodcarving studio across the yard from the giant dog. A self-taught chainsaw artist, Dennis, who had just turned seventy, makes folk-art-style wooden carvings of dozens of dog breeds. His big break came in 1995, when he and Frances sold 1,500 miniature dogs in a matter of minutes on QVC. In 2003, the couple won an Idaho tourism award for a carved exhibit about Seaman, a black Newfoundland that accompanied Lewis and Clark on their Corps of Discovery expedition. (Seaman was lucky to make it through that journey alive — the explorers ate more than 250 dogs during the trip.)

"Most people who do what I do carve bears, but I've always had a real connection to dogs," Dennis told me as he gave me a tour of his three-acre yard, which was littered with his wooden canine carvings and other funny creations — including a four-foot-long wooden toaster with a removable piece of wooden toast, and a ten-foot-tall Corning-Ware coffeepot. The yard also boasts what Dennis called the "biggest fire hydrant in Idaho." Hollow and made of concrete stucco, it was painted red and stood twice my size. During the summer months, Dennis and Frances place a portable toilet in the hydrant for the traveling public.

I didn't dare let Rezzy off the leash (we were about seventy yards from the highway), but I hoped I might snap a picture of Casey peeing on the hydrant as he explored the yard after a long stint in the RV. Instead, he took an immediate liking to the branch of a spruce tree. Casey delights in ripping apart sticks with his teeth, and while he usually appears to ingest half of what he destroys, he's somehow never had a stomach problem.

I take that back. He once had a stomach problem — but a stick didn't cause it. When

Casey was five, his midsection seemed to double in size over the course of a few days. Worried that he might bloat to death, I took him to the vet. "Has he had access to an unusually large amount of food?" she asked. I assured her he hadn't. When I got home, I studied the enormous bag of dog food that I'd left carelessly leaning against a chair on my kitchen floor. I walked around to the rear of the bag and noticed a gaping hole where Casey had been surreptitiously helping himself to dinner throughout the day.

Though it was a warm and sunny afternoon in Idaho, Dennis wore a leather jacket over a long-sleeve collared shirt. He confessed that he wasn't in the jolliest of moods. His eyes watered as he told me that his sixteen-year-old cat, Miss Tibbs, had passed away the night before. Fortunately, he and Frances still had Sprocket, a three-year-old Golden Retriever who likes hanging out in the woodcarving studio and retrieving letters from the mailbox.

"We've always had Goldens," he told me.

"Why not build a giant Golden, then?" I was curious to know.

"Well," he said, "I started by carving Beagles, so I thought it would be the surest bet. I knew I knew how to do Beagles."

Dennis suggested we check out the inside

of the dog, which is accessible by a wooden staircase near the Beagle's rear end. We entered the animal through a bright red door off the deck, which looks north toward planted fields of wheat, canola, and hay. The main sleeping area is carpeted and cozy, with a queen-size bed and a breakfast table. There's a loft in the head of the dog, and a reading alcove in the nose. The bathroom is in the dog's rear.

Dennis told me that no one has ever stolen anything from the Beagle, which he credits to man's inherent goodness. But that doesn't mean he hasn't had his fare share of taxing customers. He recalled one woman who insisted on bringing her ten Chihuahuas for a night's stay.

"You know that line about the customer always being right?" Dennis said with a chuckle. "Seldom, if ever!"

There was no Wi-Fi or TV in the Beagle, so I spent two hours relaxing on the deck under the stars. I'd been on the road for three months, and though I'd loved my journey, I was starting to yearn for a normal life again. A friend had warned me that thirteen thousand miles in an RV was an exhausting proposition; exhausting didn't come close to capturing it. I was dog-tired. I was glad to be heading east — toward home.

This was my first trip to the Potato State, and I was surprised that Idaho could compete with Oregon for greenery. Driving through the Clearwater National Forest, a verdant wilderness that stretches from the Oregon state line all the way to Montana, I passed rows and rows of towering firs as I followed the coursing Lochsa River.

Route 12, which used to be known as the Lewis and Clark Highway, stayed mostly flat, but the Idaho forest was surprisingly mountainous. (When Lewis and Clark visited the area in 1806, Clark proclaimed in his journal, "I could observe high rugged mountains in every direction as far as I could see.") It wasn't until I approached Missoula, Montana, that the horizon flattened. As I drove southeast toward Butte, the forest made way for prairie land, the road bounded by tall grass and rolling hills.

I stopped in Philipsburg, a silver-rich nineteenth-century mining town surrounded by tranquil lakes, grassy valleys, and snowcapped mountains. Philipsburg has fewer than a thousand residents, many of them cowboys and outdoorsmen who spend their downtime in one of the town's handful of saloon-style bars. But Philipsburg's most memorable building is probably its redbrick jail, where a noose hangs in clear view behind a second-story window.

I'd come to Philipsburg to meet David Riggs, a forty-five-year-old cowboy who trains Labs for a variety of roles — for example, as gun dogs and service animals for wounded veterans. We met at the Sunshine Station, a saloon/restaurant/country store/gas station where several patrons were enjoying their mid-morning shots of whiskey. David wore jeans and a cowboy hat and joked that Philipsburg is a "small drinking town with a big fishing problem." To fit in, I ordered a beer — okay, a *Corona* — with my breakfast.

The bartender spoke in a monotone and looked as if she hadn't stepped outside the dimly lit bar in years. She studied me disapprovingly, especially when David told her that I write for *The New York Times.*

"Why don't you go back to New York, then?" she said when I suggested that Montana's former lack of a speed limit might not have been the best idea.

"Be nice," David told her. "Benoit's a guest of mine."

"Well, I'm tired of people from New York making our laws," she said.

"I actually don't live in New York," I

replied. "And isn't it people in D.C. that make the laws?"

"Where do you live?" she asked.

"Massachusetts."

"That's not much better," she said.

I smiled. "And I grew up in *San Francisco.*"

She shook her head, and I decided to stick the needle in further. "I'm also *half-French.*"

That seemed to send her over the edge, because she turned away and muttered something I couldn't make out. "She's been here longer than I have," David told me when she was out of earshot, "so she has a reason to be cranky. But she's a total sweetheart underneath it all."

I got to see that for myself a few minutes later, when she returned bearing a sunny new attitude. "Do you want a lime for your Corona, honey?"

"What, now you *like* Benoit?" David asked her.

"He's all right, I guess," she said, softening before our eyes. "Poor boy can't help where he was raised."

After breakfast, I followed David to his ranch, which sits on 4,500 acres near a renowned blue-ribbon trout stream called Rock Creek. David lives there with his fiancée and ten adorable young Labs, eight of which he was training — using only positive techniques, he said — to work with wounded soldiers and children with autism. David is a service dog provider for the Wounded Warrior Project and is the founder of K9 Care of Montana, a charity providing customized outdoor activities with dogs for children with special needs.

That afternoon, we took Casey, Rezzy, and David's four-year-old yellow Lab, Rocky, for a swim in the East Fork Reservoir. Rezzy had learned to swim in Oregon by following Casey's lead as he bounded into several lakes; she loved racing him for a tennis ball far out into the water. Watching Rezzy was a reminder of just how much dogs learn by observing each other.

While the animals played, I asked David about some of the commands he teaches dogs that are eventually paired with injured vets. He told me about one, called "Watch Your Six," which prompts a dog to stand or sit behind a veteran.

"There's something about having an extra pair of eyes behind you that gives comfort to wounded vets with post-traumatic stress," he explained. "So many of these guys come back from Afghanistan paranoid as hell, and having the dog behind them offers a real sense of security."

David also told me about another directive, "Hold," which he was teaching to a dog that would be paired with a 275-pound wounded vet with a spinal injury. Should the vet fall, the "Hold" command prompts the dog to position itself under the man's armpit and push up, providing brace support and helping him off the ground.

Later that night, David and I sat outside his house by a roaring fire. It's difficult not to get philosophical under the stars on a Montana ranch, and before long David and I were chatting like the oldest of friends.

"Man, I came here to get away from the world," he said. "I'd had enough and wanted to be alone."

(Montana is certainly a good place for that. In *Travels with Charley,* Steinbeck wrote that "the frantic bustle of America was not in Montana." Steinbeck could not get enough of the state. "I am in love with Montana. For other states I have admiration, respect, recognition, even some affection, but with Montana it is love.")

David continued: "Man, the dogs saved me. I wanted to be left to myself, but the dogs brought me back into society. Being a dog trainer got me out of myself and my head."

I asked David about his previous yellow

Lab, Hopper, who died a year before my visit. Hopper used to retrieve just about anything David asked him to. As a joke, David once had the dog snatch a bag of cash from a temporarily open safe behind the bar at the Sunshine Station.

"In a lot of ways I still haven't gotten over losing him," he told me. "Hopper was the best dog I've ever had. I got him at five weeks, and he would sleep on my chest from the beginning. At eight weeks, he was already retrieving birds. As an adult dog, he was shockingly smart. It was like he would think before he acted."

Hopper's father, Cropper, was the first yellow Lab in the history of the AKC to win the national title in Field, which tests hunting dogs on a variety of tasks. And while Hopper was a great hunting dog like his father, his utility to David went further. David walks with a heavy limp as a result of a spinal cord injury and numerous surgeries, and he relies on his primary dog to help him up when he falls.

David told me about the time he rolled his SUV over the side of a hill outside Philipsburg, ejecting Hopper from the vehicle. When the truck settled, the dog jumped back inside; David awoke to Hopper licking blood off his face. They climbed out of the truck together, and David crawled up a hill and collapsed in the middle of a road. When two women came upon the pair, Hopper was lying with his chest on top of David.

"He was hesitant to allow the girls close until he realized they were there to help," David told me. While David was transported to Missoula with a broken hip, shoulder, and ribs, Hopper was taken to a local vet and received twelve stitches to the shoulder.

After thirty minutes by the fire, I asked David if Hopper was the dog of his dreams.

"Absolutely," he said. "I think Hopper was so special because he was my service dog, my hunting retriever, and my companion. But a dog is really only as good as the experiences you put him through over time. Eventually the dog reflects your personality and your past, because he's alongside you for a decade or more of your life."

He looked down at Rezzy, who was seated at my feet at the end of a leash, peering intently into the darkness of a nearby field. "She's always observing, listening, seeing what animal might be out there," David said. "I have to say, Rezzy's an amazing animal to me."

"And to think I considered not taking her from that reservation," I told him.

David struggled to stand up from his

chair. "I believe things happen for a reason," he said when he was firmly on his feet. "It's evident to me that she was meant for you, at this time in your life." He started hobbling back toward his front door, shuffling through the dirt. Then I heard him stop. After a few beats of silence, I turned around in my seat to see what he was up to. But I couldn't place him in the dark.

"Are you okay?" I asked.

"Yeah, yeah, I'm good," he assured me. "I was just thinking whether you were meant to save Rezzy, or whether she was meant to save you and Casey. In the end, I guess that's for you to figure out."

The drive south from Philipsburg to Yellowstone the next morning was sufficiently stunning that I wondered if the famed national park would even faze me. Heading south on Montana Route 1, I passed the impressive Georgetown Lake, which was separated from the road by just a thin strip of grass and the occasional veil of trees. Further south, the Montana scenery wasn't quite so green, but the miles of gold grass were still a striking contrast to the bright blue sky.

I was thankful to have a personal tour guide for my first trip to Yellowstone — a friendly and animated thirty-six-year-old park employee by the name of Kevin Franken. (A mutual friend had put me in touch with him.) Kevin had the day off and was kind enough to chauffeur me around the nearly 3,500-square-mile park in his civilian car, a blue metallic 1994 Buick Park Avenue.

Not long into our tour, I asked him how he became a park ranger. "Funny story about that," he said. "Originally I was going to be a lawyer; I went to law school and everything. But then I took one of those online personality tests that tell you what career you should go into."

"And it said you should be a park ranger?"

"It did! And I was, like, 'Duh, why didn't I think of that?' I love nature, love interacting with people. And the rest is history. I've worked in four national parks since."

Near Mammoth Hot Springs, a grouping of geysers near one entrance to Yellowstone, Kevin pointed out a lone male bison in the distance. "Last year, we had a bunch of bison give birth right on the front lawn of our ranger office building," he recalled. "Adult bison are brown, but their babies are an orange-red color and don't really look like bison. So we had a few tourists come up to us and ask a very silly question:

'What are those strange-looking orange dogs hanging out with the bison?' "

"They thought the baby bison were dogs?"

"They did. I guess they assumed they were a breed of dog specific to Yellowstone or something. We get all kinds of silly questions from tourists."

"Like what?" I asked.

"One of my favorites is, 'What time do the rangers make Old Faithful go off?' As if we're controlling it by remote control. There's a whole book of ridiculous things tourists say to park rangers."

Later, as we passed the Hoodoos rock formation, I complained that the park isn't very dog-friendly. Dogs aren't allowed — even leashed — in the backcountry or on trails or boardwalks. I'd grudgingly left Casey and Rezzy in the Chalet.

Some fifty years before my trip to Yellowstone, John Steinbeck also complained about the park's then dog policy. When a park employee told Steinbeck that Charley would have to be leashed in the park on account of "the bears," Steinbeck wrote that he told the ranger, "This is a dog of peace and tranquility. I suggest that the greatest danger to your bears will be pique at being ignored by Charley."

I asked Kevin if many dog-loving visitors

to Yellowstone complain about not being able to bring their dogs on trails. "We get some who do, but most people realize that this is a *wild* area — and they don't want to see their dogs killed by bears, bison, or elk," he said.

"I guess that makes sense," I conceded. "And Rezzy does like to chase things."

"Many dogs do. It's just safer for everyone this way. I've heard stories of eagles swooping down and snatching small dogs. And a coyote killed an off-leash dog a few years ago."

There's another reason to outlaw dogs from much of the park: water-loving dogs sometimes end up in scalding geysers. The book *Death in Yellowstone: Accidents and Foolhardiness in the First National Park* tells the tragic tale of a large dog, Moosie, who jumped into the park's Celestine Pool (and its 202-degree water) on a summer day in 1981.

As a friend of the dog's owner prepared to leap in after the animal, someone screamed, "Don't go in there!"

"Like hell I won't!" the man responded before diving headfirst into the hot spring. The story doesn't end well. The man never did rescue Moosie, and when he made it back to land, he had third-degree burns over his entire body. He died the next day.

"It's a sad story, and I really don't like to tell it — especially to dog lovers," Kevin said.

He was much more willing to talk about wolves, which were reintroduced to Yellowstone in 1995 and feed on young, old, or wounded elk. But I turned the conversation to bears. I'd never seen one outside a zoo, and I was intent on changing that.

"People think they see grizzlies everywhere," Kevin told me. "They'll stop in the middle of the road, get out, point at a boulder in the distance, and say, 'Oh my God, it's a bear!' And I'm, like, 'Nope, that's a rock.' "

But there are plenty of bears in Yellowstone — some six hundred grizzlies alone, in fact. The summer before my visit, an adult female grizzly was killed by the National Park Service after DNA evidence placed the bear at the location of two fatal maulings of hikers. (The deaths were the first caused by a Yellowstone grizzly in twenty-five years.)

Kevin had just finished telling me about the killings when, lo and behold, we spotted a grizzly in a grassy clearing not far from the Ranger Museum. Kevin pulled his Buick to the side of the road. We got out, and I started filming the bear with my iPhone. The animal appeared to be making its way toward us, and when it was about sixty yards away, Kevin surmised that it was hoping to cross the road at precisely our location.

When it was within forty yards, Kevin suggested we get back in the car. He didn't need to tell me twice; I'd already started backpedaling toward the passenger-side door. Safely in the Buick, I kept my iPhone video rolling as the medium-sized bear sauntered across the road, some ten yards from Kevin's windshield.

"That was so cool!" I said soon after, doing my best impression of an easily impressed ten-year-old.

Still, part of me was sad that Casey and Rezzy missed out on the bear. My dogs were my travel companions; Casey, in particular, had been at my side for nearly every memorable experience of my journey. It felt wrong when my dog missed something — almost like it never even happened.

On my way to Mount Rushmore the next day, I pulled over for gas on the North Cheyenne Reservation in Lame Deer, Montana.

Since rescuing Rezzy, I'd made a habit of

stopping on Indian reservations and count-
ing the stray dogs hanging around gas stations and convenience stores. Most times, I
would strike up conversations with locals
about the dogs. Though some of the animals
had owners who let them roam, most were
strays.

As I fed and played with the handful of
dogs at the Cheyenne Depot gas station,
Rezzy sat in the passenger seat of the Chalet
and watched the action. I wondered what
was going through her mind. Did gas stations have any special significance for her?
Did she have a sense this was a reservation
similar to the one where she'd lived? Where
was a pet psychic when I needed one?

It was never easy to leave rez dogs behind,
but I got back on the road and drove for
what felt like forever on U.S. 212 in southeastern Montana. Sometime after I crossed
into South Dakota, the plains became hillier
and pines once again lined either side of the
road. As I approached Mount Rushmore,
the highway sloped upward, guiding the
motorhome up from the Badlands toward
those four famous faces carved into rock.

I'd never been to Mount Rushmore, and I
was annoyed — for the second time in as
many days — that dogs weren't allowed
where I wanted to go. I'd hoped to take a

460

picture of Casey and Rezzy with the presidents in the background, but the closest I could get them to Abraham Lincoln was a small area near the parking lot. So I left the dogs in the RV and walked to the memorial by myself, navigating a few hundred of the nearly three million tourists who visit the monument each year. I made my way to the viewing area and gazed at the presidents, all of whom — except Jefferson — appear to be mildly irritated.

I was mildly irritated, too. I wasn't especially impressed by Mount Rushmore, and I had a daunting stretch of highway to complete over the next two days — some seven hundred miles to Kansas City. My rear had started hurting again in the mountain West, and after months on the road I was sick of every song on my iPhone. My Facebook followers had tried to help by suggesting dozens of new songs, from canine-themed country ballads ("Like My Dog" by Billy Currington) to classic rock ("Runnin' Down a Dream" by Tom Petty and the Heartbreakers) to goofy, old-school hip-hop ("Summertime" by DJ Jazzy Jeff and the Fresh Prince).

My mom called from France as I was walking back to the Chalet from the viewing area. "Where are you?" she wanted to know.

"Mount Rushmore," I said.

"Oh, better you than me!"

"I know. You'd hate it — so many tourists."

"Aren't there better places you could go?" she asked.

"I was in the area," I told her. "And I wanted to get a picture of the dogs in front of the presidents, but dogs aren't allowed. It's not like France, where dogs can go just about everywhere."

"Oui, ça c'est vrai," she said, switching mid-conversation — as she often does — to French.

"But why don't the French pick up after their dogs? I remember when you'd take me to Paris as a kid, there would be poop everywhere on the sidewalk."

"Tradition, I suppose," she said. Then she changed the subject. "You sound tired."

I'd never followed through on my promise to meditate, and I was hoping she wouldn't bring it up.

"Driving this much is really hard," I told her. "I'm not sure how much more I can take in this RV before I lose my mind."

"I've found that meditation really helps in stressful moments," she said.

462

I'd walked right into that one.

Before letting my mom go, I decided to tell her about my trip to see Dr. Gold in Manhattan. I'm not sure why I chose to open up to her in the Mount Rushmore parking lot — maybe the fatigue was getting to me.

"In New York City, I went to see a therapist who specializes in the human-animal bond," I began, pacing around the RV as I spoke.

"I didn't know such a thing existed," she said.

"It does," I confirmed. "I told him about some of my issues with Casey, stuff that I've already mentioned to you. But he had an interesting theory. He said that I was expecting Casey to make up for the love and nurturing you didn't always show me as a child."

There was silence on the line. My mom and I had already had many emotional talks — both on the phone and in person — about my childhood, but bringing Casey into the mix was a new wrinkle. I could imagine her thinking, *Oh, so you're blaming me for your dog problems, too? Might there be something that's not my fault?*

If she thought that, she didn't let on. Instead, she asked me to elaborate. Specifically, she wanted to know if my feelings had changed. Had there been some resolution?

"Yes, and it's a good thing," I said, before realizing that maybe she meant a resolution of my feelings *toward her.* But I wondered if she'd stumbled onto something; maybe fixing one problem (my insecurities around my dog) helped me fix the other (lingering anger toward my mom).

I breathed deep and tried to access the wisdom stored in my gut, as Dr. Gold had suggested at the start of my journey. I couldn't locate any insecurity or anger there. Maybe, I thought, I was just too tired. Or maybe I'd succeeded in killing two psychological birds with one long road trip.

To get to Kansas City, I drove across South Dakota on Interstate 90, an interminable stretch of road that won't win any highway beauty pageants. I passed the time by calling Marc and then Dylan, my friend who'd worried that I wouldn't make it back to Provincetown in one piece.

"I don't want to jinx myself," I told him, "but I haven't crashed this thing yet!"

After spending the night at the KOA in Sioux Falls, I picked up a hitchhiker and his dog on my way south on I-29 toward Omaha. I'd passed dozens of hitchhikers on my journey, most of whom looked like prison escapees. I'd hoped to get lucky and come upon film director John Waters, who was hitchhiking across the country for a book at the same time as I was driving across America. But no such luck.

Instead, I stopped for a middle-aged man who looked remarkably like Harrison Ford, albeit after a few days without bathing. At first I thought it might actually be the actor, which is probably why I stopped. (That, and he had a dog.) It wasn't, and the man's fluffy little terrier mix turned out to be a nightmare. The dog — named Hyper — wouldn't stop running around the inside of the Chalet and nipping at Casey and Rezzy, who stared at the animal as if it were some kind of insane canine space alien.

The Harrison Ford look-alike, whose name was Jerry, proved to be equally strange, mostly mumbling one-word answers to my attempts at conversation. I gathered that he was on his way to Iowa, where he planned to meet a woman he'd fallen in love with on the Internet, and who had recently left her husband. Jerry said he'd spent the past year in North Dakota, working odd jobs and spending his free time on a public computer at a local library, sending love emails to the woman.

During the nearly three hours we spent together before I dropped him off north of Omaha (so that he could catch I-80 toward Iowa), Jerry didn't ask a single question of me. So I filled as much dead air as I could. When I asked about Hyper, he told me he'd found the stray peeing behind a snowbank, and that Hyper had interpreted his passing interest as an invitation never to leave his side.

"I'm not really a dog person," Jerry said, "but he wouldn't let me be."

Hyper had followed Jerry back to his one-room basement apartment and then refused to be shooed away. "I closed the door and opened it three hours later, and he was still there!" Jerry told me in a momentary burst of personality. "I felt bad, so I let him inside."

The dog would sprint in circles around Jerry's apartment for much of the day, finally collapsing on his bed at night and sleeping until about noon. When Jerry decided to follow his heart to Iowa, the woman he was going to meet insisted he bring Hyper with him.

"She's a dog person," he said sheepishly.

"Sounds like you're stuck with Hyper for good," I told him.

"Yeah," he replied, staring straight ahead.

We drove in silence for miles before I asked him if anyone had ever told him he looked like Harrison Ford.

"All the time," he said.

I waited a few seconds to see if he might go into more depth. Nothing. "Does that ever get annoying?" I asked.

"All the time," he repeated.

"Am I annoying you now by bringing it up?"

"Not really," he said.

"Have you ever pretended to be him?"

"Harrison Ford?"

"Yeah."

"Once."

"What happened?"

"I was shit-faced drunk. My friend bet me money that I couldn't convince a girl that I was Harrison Ford."

"Who won the bet?" I asked.

"My friend."

"Why? You weren't convincing?"

"I guess not."

Another few miles of silence.

Then, unprovoked: "You should really clean your windshield," he said.

In Kansas City, I stayed overnight at the home of a woman who'd been following my journey on Facebook. She'd generously of-

467

fered the dogs and me her family's guest bedroom — to the chagrin of her husband, who wasn't so sure about a stranger from the Internet squatting in their home.

On my first full day in Kansas City, she drove me across town to Rose Brooks, a domestic violence shelter that was in the process of building an on-site kennel for dogs. There I met a woman named McKenzie and her 140-pound Great Dane, Hank. Though technically a dog, Hank looked more like a slender horse.

McKenzie and Hank had stayed at Rose Brooks the previous year, after her ex-boyfriend threw her through a wall and tried to beat her with a hammer. Hank had saved McKenzie's life by stretching out on top of her, but the ex-boyfriend then pushed Hank off the porch, breaking his hip and ribs. When McKenzie called Rose Brooks, the shelter initially said she couldn't bring her dog.

"I was like, 'Nope, we're a package deal,' " McKenzie recalled in a conference room at the shelter. Her abuser had gotten out on bail soon after the attack, and she worried that he would try to hurt Hank. "He knew that this dog means more to me than anything in the world."

Concerned that McKenzie's ex would try

to kill her, the shelter relented on its no-pets policy and made a space for the dog in a basement bathroom. Before long, everyone at Rose Brooks loved Hank, and the staff noticed that his presence brought smiles to the faces of the other women and children at the shelter. Combined with the knowledge that many battered women won't go to a shelter if it means leaving their pets behind, Rose Brooks decided to build a kennel on its property.

Frank Ascione, a developmental psychologist at the University of Denver who researches violence against animals, has found that 71 percent of domestic violence victims had their pets killed, harmed, or threatened by their abuser. "Abuse directed against animals is indisputably linked to child maltreatment and domestic violence," Ascione and Phil Arkow write in *Child Abuse, Domestic Violence, and Animal Abuse.*

Though men are more likely than women to abuse animals, there are exceptions. In an examination of thousands of cases reported to Pet-Abuse.com by Kathleen Gerbasi, she found that women are more likely to hoard pets (a form of abuse in many cases) and almost as likely as men to neglect or abandon them.

Will kids and teenagers who abuse pets

become violent adults? Though that link is often assumed, the connection between childhood animal cruelty and later criminality is tenuous. What's clear is that many violent adults have no history of hurting animals. School shooters, for example, aren't usually the vicious animal abusers the media can make them out to be. A 2002 task force organized by the Department of Education and the U.S. Secret Service found that of forty-one attackers in school shootings, only five had such a history.

Nor is there any real evidence that youthful animal mistreatment is strongly correlated with adult violence. "The awkward fact is that most wanton animal cruelty is not perpetrated by inherently bad kids but by normal children who will eventually grow up to be good citizens," Hal Herzog points out in *Some We Love, Some We Hate, Some We Eat.*

Several studies of college students have found that roughly half admit to either perpetrating or witnessing acts of cruelty to animals. Even Charles Darwin admitted to once beating a puppy, "I believe simply from enjoying the power," he wrote. "This act lay heavily on my conscience, as is shown by my remembering the exact spot where the crime was committed."

Though Gandhi is credited with saying that "the greatness of a nation and its moral progress can be judged by the way its animals are treated," he must have forgotten about Nazi Germany, which in 1933 enacted extraordinary legislation to protect animals from suffering. Dogs at the time could not have their ears or tails docked without anesthesia, and no animals could be treated inhumanely in scientific experiments or in the production of films.

"I will commit to concentration camps those who still think that they can continue to treat animals as inanimate property," Nazi Germany's Hermann Göring said. "To the German, animals are not merely creatures in the organic sense, but creatures who lead their own lives and who are endowed with perceptive facilities, who feel pain and experience joy and prove to be faithful and attached."

Today, even as most states in this country have beefed up their animal cruelty laws, animal control officers struggle to define animal abuse. Anticruelty codes specify that animals shouldn't be deliberately mistreated, but the general public — and even officers themselves — don't always agree on what constitutes cruelty. Is a family in agrarian California that loves its dog but leaves it alone in the yard, sometimes for days at a time, abusing their dog? What about a debutante who buys her dog fancy sweaters and dog beds but takes the animal outside only for one brief walk each day?

Further complicating efforts to prove cruelty to animals is the problem of gathering evidence. Often, the best — or only — witness to the abuse of a human is the victim. But animals obviously can't report their abusers. Animal control officers routinely drop cruelty cases because of a lack of evidence, leaving helpless dogs with owners who the officers believe (but can't prove) to be cruel.

During my months on the road, I'd heard about Americans mistreating their dogs in gruesomely diverse ways. There was the Montana man who gave his twenty-pound Pomeranian a "to-go cup of Vodka" (the dog couldn't stand up and was four times over the legal driving limit for humans), the North Carolina man who dragged his Labrador/Chow mix behind a moped, the New Jersey woman who attempted at-home surgery on her dog (she also gave the animal antidepressants meant for humans), the two California men who lit fireworks strapped to the back of a Dachshund, and the St. Louis man who photographed himself having sex with his dog.

Next to killing a dog, bestiality may be the most reviled form of canine abuse. But because of its highly taboo nature, we don't know much about its prevalence. The famous *Kinsey Reports* from the 1940s still provide the most comprehensive data: Kinsey found that 8 percent of men and 3.6 percent of women had engaged in sexual contact with an animal. (He estimated that the numbers were higher in rural communities.)

After nearly two days in Kansas City, I shooed Casey and Rezzy into the RV and got back on the road. It was a windy May morning in Missouri, and it was nearly impossible to keep the RV in my lane as I drove east on Interstate 70 toward St. Louis. I was blasting the Crash Test Dummies song "Mmm Mmm Mmm Mmm" when I heard the siren behind me.

The officer took his time getting out of the cruiser, and when he finally asked me if I knew why he'd stopped me, I told him that keeping the RV steady in these conditions was about as unlikely as the hapless Kansas City Royals winning a World Series. It was a miracle I hadn't been pulled over earlier in my journey.

He broke into a half-smile and asked where I was heading. I pointed to Casey and Rezzy — wedged uncomfortably in their usual spot between the seats — and told him about my journey and my book.

"Oh, like *Travels with Charley*?" he said. "Why don't you come back to my car and meet my dog?"

"You're a K9 cop?" I asked. I couldn't believe it. The next day, I was planning to ride along with a K9 officer in Belleville, Illinois, just over the state line from St. Louis.

I carefully exited the Chalet's side entrance (the furthest door from the big rigs barreling past us) and made my way to the officer's cruiser. There, I stood awkwardly outside his passenger-side door, waiting for confirmation that I should get inside.

"I'm not arresting you — get in!" he said with a laugh.

The officer's dog, a large German Shepherd, lounged behind a grate in the rear of the car. "He's a great dog," the officer said, "but sometimes he can be a pain in the ass, just like any other dog."

The officer seemed happy to have someone to talk to. He lamented that many cities and towns across the country had slashed their budgets in recent years, meaning that fewer dogs were on duty.

"These dogs can do things we just can't," he told me. "They're faster than we are, they can get in places we can't, and they have better noses." (Rapper Snoop Dogg learned that the hard way in 2012 while on tour in Norway, when an airport police dog marked Snoop as carrying marijuana as soon as he walked into the terminal.)

Dogs have been used to patrol the streets and control people since the Legions of ancient Rome. They were especially popular on the night watch in Belgium, where the first training school for police dogs was established in 1899. Neighboring Germany followed suit in the early twentieth century, developing new breeds trained for increasingly specialized police work, though some in law enforcement questioned whether the animals added much value.

Dogs were almost removed from New York's police force in 1920, when the police commissioner at the time claimed the dogs didn't actually make the city safer. The city's police dogs back then would roam neighborhoods with detached houses, barking when they spotted possible evidence of a burglary (an open ground-floor window, for example) or a sign of danger (smoke coming from a window). In retrospect, that probably wasn't the smartest use of a dog's skill set.

In 1971, the NYPD procured a bomb-sniffing dog, Brady, who, *The New York Times* wrote, was "viewed skeptically until she detected explosives on a plane at Kennedy Airport." Today, police dogs carry out a wide variety of critical tasks. They chase after suspects, search for lost people, and sniff out everything from narcotics and explosives to human remains. Specially trained dogs also accompany the U.S. military on its most challenging missions. In 2011, a Belgian Malinois named Cairo was part of the Navy SEAL team that killed Osama bin Laden. During the Iraq War, so many U.S.-trained combat dogs were injured that the army opened a surgical care unit for dogs.

It felt strange to sit in a squad car on the side of a highway talking about dogs. Though I'd done nothing wrong, I never got comfortable in the cruiser and never thought to ask for the officer's name. Eventually, he said I could go. I was halfway back to the motorhome when I realized I'd forgotten to tell him something. I hurried back to his cruiser and leaned in the passenger-side window.

"If you want," I said, "you can follow my trip on Facebook."

■ ■ ■ ■

The citizens of Belleville, Illinois, were (unfortunately) well behaved the next day during my drive-along with officer Brian Dowdy and his two dogs — a nine-year-old German Shepherd, Steve, and a four-year-old Belgian Malinois, Art.

It was midday on a Saturday, and the police dispatcher wasn't giving us much to work with: a barking dog, a stolen cell phone, a fender-bender.

"Does no one commit crimes in this town?" I asked, eager for some action.

"It's a Saturday morning," he said. "The bad guys are still sleeping."

As we drove around this city of 44,000, Brian, who was friendly and handsome and looked younger than his thirty-seven years, told me that he'd always wanted to be a K9 officer.

"I had sixteen dogs growing up," he said, "and I only wanted to be a cop if I could have a dog with me."

Brian began his crime-fighting career as a bike cop for a hospital, where he lobbied unsuccessfully for a canine sidekick.

He now runs his own police dog training school and never goes on patrol in his SUV without a dog. Brian's police vest is equipped with a button that opens a hydraulic door on the back of his cruiser, in case he's in a tight spot and needs his dog to come to the rescue.

"When that door opens," Brian explained, "my dogs know they're supposed to bite something."

I asked him what he looked for in a police dog. "Personally, I like a dog with a tremendous amount of confidence," he said, sipping on a Mountain Dew and gnawing on some Kodiak chewing tobacco. Brian couldn't get over the fact that I didn't like Mountain Dew, which he considers the world's greatest beverage. "I like dominant dogs, Alpha dogs, dogs that will want to challenge me occasionally. Because when they get out in the field, I've learned those dogs have no trouble challenging a bad guy. Not every officer wants that, but for me it works."

Typically, Brian has only one dog at a time in his cruiser. But we were on our way to a town fair where Brian had been asked to give a public demonstration about police dogs, and he wanted to use both animals. Art was fairly new to the job.

"I've only had him for a year," Brian said, "and he's come a long way in that time. For the first few days he was in my car, I couldn't get him out."

"Why is that?" I asked.

"Because he'd try to bite me in the face," Brian said with a chuckle.

At that moment, Art started barking at a car that had pulled up next to us at a light.

"What's he riled up about?" I wondered.

"The guy in that car's looking at him," Brian explained. "If they don't look at him, he's fine."

Art's bite command is "Attack," and Brian didn't need to tell the dog twice during the demonstration at the town fair. When Brian said the magic word, Art catapulted himself at an officer wearing a protective jacket, biting down on the man's arm and refusing to let go. Brian told the crowd that Art was born in France and preferred receiving his commands in French.

"Aux pieds!" Brian said a few minutes later, prompting the dog to heel.

When someone in the crowd asked if Art lived with Brian, he confessed to letting the dog sleep in bed with him. "He's spoiled," Brian said. "I try not to give him people-food, but when he's looking at you with those *eyes,* it's hard not to." (Brian was further proof that even many of those who live with working dogs — and who know,

479

intellectually, that they shouldn't spoil their animal — are as powerless as the rest of us when faced with a dog's pleading gaze.)

Next, it was Steve's turn to try to remove a chunk of the fellow officer's arm. Brian told the crowd that Steve almost never had the chance to apprehend fleeing suspects.

"He was supposedly vicious and untrainable, and the people who had him were going to put him down," he said. "So I took him, because I saw his potential." Steve ended up being the best dog Brian has ever worked with. "He's vicious when it comes to bad guys," he told some young teens seated in the first row of spectators. "But if you're a good guy, then he's a lover, not a fighter."

Brian further impressed the teens by demonstrating that to get Steve to bark, all he had to say were two words: "free beer!" Steve barked ferociously as Brian screamed, "Free beer! Free beer! Free beer!"

A police dog's bark is usually enough to end a fight. Earlier that day, Brian had recounted the time he responded to a bar brawl involving dozens of people. When Brian and Steve arrived, there were ten cops on the scene struggling to disperse the crowd.

"I said 'free beer,' Steve started barking,

and everyone started behaving themselves real quickly," he recalled. "Sometimes, having a dog is like having a gun."

But not all criminals throw up their hands at the sight of a police dog. Later in the day, as we resumed our patrol, Brian told me about the many household appliances — pots, pans, televisions — that fleeing suspects have hurled at his dogs.

"One guy even threw a cat," Brian told me as we circled a rundown hotel where he was on the lookout for drug activity. "You can imagine how that turned out."

"What happened?" I asked, fearing the worst for the poor animal.

"What do you think happened? The dog killed the cat. He caught it in the air, shook it once, and it was dead."

Brian spent a good portion of our time together apologizing for the lack of crime in Belleville that day, and I wondered if he might start inventing reasons to pull people over. Our first target was a woman driving a silver Pontiac Grand Am with a busted taillight and a bumper sticker that read, "Ask Me If I Give A Fuck."

During our second stop, of a dark Honda driven by a man with prior drug convictions, I finally got to see Art in action. He circled the car and alerted Brian to the presence of drugs inside by sitting down near the passenger-side door. (Brian trains his dogs to sit if they smell drugs.) But a further search of the vehicle by Brian revealed only cannabis residue.

"If there's anyone else you want me to pull over, just say the word," Brian joked as his shift was coming to an end.

I asked Brian if there was anything else I should know about police dogs. He took a swig of Mountain Dew and looked in the rearview window at his animals.

"These dogs are really important for our image — the way the public perceives us," he said. "People are much more likely to talk to me when I'm in uniform if I have my dog with me. Having a dog humanizes me. Dogs stimulate conversations with the public. People stop and ask about them, kids want to meet them. The dogs make us more approachable. Suddenly we're not the *big bad police* — we're just another guy who loves dogs."

And Brian loves dogs so much, he said, that he'd likely quit the force if he couldn't ride with one.

"I don't really want to have any job where I'm not allowed to bring my dog to work," he said. "Know what I mean?"

"I do," I replied, thinking of my dogs napping at my feet every day as I write.

Two miles from St. Louis's soaring Gateway Arch, across the Mississippi River, there's a forgotten city named East St. Louis.

A notorious example of inner-city blight, it's an "urban prairie" of abandoned and burned homes, boarded-up former businesses, decaying streets strewn with garbage, and entire blocks overrun with vegetation. I've visited many ghettos in my work as a journalist; East St. Louis is the bleakest.

It is also the deadliest. This city of 27,000 has the highest per capita murder rate in the country (seventeen times the national average), and every few blocks a stop sign or other makeshift shrine is decorated with clothes or teddy bears, signifying where someone has been murdered.

Amidst the wasteland, dogs eke out a hardscrabble existence. They roam the city's potholed streets, sleep on discarded mattresses in derelict buildings, and hide out in overgrown fields. These "wild dogs," as writer Melinda Roth calls them, are "genetic castaways on the third rail between domesticity and wildness."

The dogs of East St. Louis have it worse than rez dogs. On Indian reservations, drunk teenagers sometimes try to run over dogs with their cars. In East St. Louis, sober teenagers shoot dogs for sport. Many dogs are trained for fighting, or used as bait to test the bloodthirstiness and fighting instincts of promising Pit Bulls. Even the dogs of what are considered responsible owners in East St. Louis tend to spend their lives chained to a garage or fence post, circling the same ten square feet of dirt or broken pavement.

I'd come here to spend two days with my friend Randy Grim, the fifty-year-old founder of Stray Rescue of St. Louis. I'd first learned of Randy when I picked up his book *Don't Dump the Dog,* which is part dog training manual and part diary of a neurotic, self-deprecating dog rescuer. Randy is foulmouthed, germaphobic, and socially anxious. The only living beings that don't scare him are dogs.

"I've hosted parties I did not attend," he confesses in *Don't Dump the Dog.* "I flush *my own* toilet with my foot. During the course of my lifetime, I have visited the emergency room with symptoms of the plague, botulism, bird flu, tuberculosis, West Nile, malaria, and anthrax poisoning."

Randy's also terrified of bridges, which I discovered when I spent a week rescuing dogs with him in 2009. On a cold, overcast

484

December morning, he'd driven us around in his white Jeep, which was strewn with dog bowls, dog food, leashes, cigarette cartons, and bottles of hand sanitizer. As we rolled over the Martin Luther King Bridge, he blasted the Ramones' "I Wanna Be Sedated." Then he grabbed my wrist with his free hand, looked at me with abject terror, and screamed, "Oh my God, we're going to DIE IN THIS CAGE!"

He was fine once we were safely on the other side. "As you can see, I have issues," he said, lighting his third cigarette in fifteen minutes.

It probably takes a man with issues to drive — unarmed — around the most dangerous city in America, rescuing dogs and trying to convince gang-bangers to treat animals with something approximating respect. Randy has been carjacked, chased from backyards and abandoned homes, and threatened with a gun.

"That's the guy who almost shot me," Randy told me soon after the bridge incident, pointing to a man milling around outside a small brick house. Randy smiled and slowed the car to a stop, as if he wanted to say hello.

"Are we really stopping here?" I asked. It was my turn to look terrified.

485

"Oh, he won't shoot us," Randy assured me. "He and I are buds now. He just didn't understand at first that all I wanted to do was feed his dog."

"What's up, homie?" the man said as we got out of the Jeep. He was beaming at the sight of Randy, who wore a baggy sweatshirt and jeans over his tall, soft-featured frame. "You must have known my dog was hungry!"

"I was just telling my friend here how we met," Randy said.

"Oh, yeah, that was a trip."

"You thought I was a crazy white guy!"

"Yes, yes I did. I was like, 'What's this crazy white dude doing snooping around?'" He turned to face me. "At first I thought Randy was a cop. Then I thought he was on dope."

"If I remember correctly, your exact words were, 'I'm going to pop you,'" Randy reminded him. "So I said, 'Well, if you're going to kill me, can I at least feed your dog first?' You didn't know what to say to that."

"I'm a changed man now, though," the man insisted. "I've found God."

"And thank God for that," Randy said.

Later that day, as we drove past a yard with a Rottweiler chained to a crooked, ramshackle doghouse (a sign on it read *Be-*

ware — Crazy), I asked Randy why he never carries a weapon.

"Because I'd be on Death Row," he said simply. "I know myself. I'd get so angry at the way people treat their dogs out here, or at the people from the suburbs who come to dump their dogs here when they don't want them anymore, I'd go on a rampage. And I'd never forgive myself if a dog got killed in the crossfire."

I'll never forget that first week with Randy. Together, we found and rescued eight puppies from an abandoned building and a sweet, emaciated Pit Bull from behind a graffitied house; we convinced a crack addict to help us corral two more young strays, one of which Randy named Benoit; we lured a very pregnant stray into a humane trap and rushed her to a vet; and we poured bags of dog food for several feral packs that wouldn't get within thirty yards of us. (Feral dogs are usually born on the streets and don't have much contact with humans.)

Driving around East St. Louis with Randy was also heartbreaking. First, there was the human suffering; I'd never seen poverty like this in America. *Time* profiled East St. Louis in 1969 and proclaimed that "only the brave dare walk its streets after dark." It's only gotten worse since then. I thought about what it must be like to grow up here, smack-dab in the middle of hell on earth.

"When I was in New Orleans rescuing dogs after Hurricane Katrina, most of the rescuers who'd come from around the country to help couldn't get over how depressing the Ninth Ward was," Randy told me. "But it's Katrina *every day* here. And no one cares. If society doesn't care about the people here, it sure as hell isn't going to do anything about the dogs. As bad as the people have it, the dogs have it worse."

"I suppose people who are beaten down need something to beat down, too," I said.

"Exactly. But I still hate people for it. The dogs walk around here with limps, cuts, gunshot wounds. And no one gives a shit."

Just about every house or backyard we passed had a dog's death associated with it. "I've found dead dogs in that house, and some in that one, and some over there in that one with the overturned couch in the driveway," Randy said as we rolled down yet another street littered with garbage and rusted-out furniture. "In the winter, so many puppies freeze to death in these houses."

A few minutes later, Randy and one of his rescuing partners, Donna, convened at an abandoned one-story house that had recently been destroyed by a fire.

"So many dogs I loved died in there," Randy said, facing the charred, white-shingled structure. "For some reason, they all came here to die."

Glad to see the house gone, Donna and Randy held hands and skipped up to what remained of the building. They walked gingerly through its remains, joking about how they might decorate the place.

"I think this area would make a nice study," Randy said.

"I'll take this bedroom," Donna added.

When they'd finished celebrating the house's demise, Randy screamed, "Good-bye, fucked-up death house!"

According to Alan Beck, who wrote *The Ecology of Stray Dogs,* the average life expectancy of a stray is just over two years.

"Dogs out here look like they're ninety-five, but they're actually only one or two," Randy explained. "It's amazing to see the transformation once we get them fixed up and cleaned up and in foster care or adopted. Without the stress of living out here, they look their age again."

Over the course of our first week together, Randy and I came upon a handful of dogs he wanted to save but couldn't. Either he didn't have space at the Stray Rescue of St. Louis shelter, or the dogs weren't in bad enough shape.

"I'm basically like a vet and a dog psychologist," he explained. "I'm triaging out here. I look for the dogs who need medical attention first. If it's a close call, then I try to read the dog's mind."

"What do you mean?" I asked.

"I mean, are they happy enough out here? Will they survive if I can only rescue them in a week or two? The problem with a lot of these dogs is that they don't really know how to be wild dogs. You'll see a starving dog walking down the street with a dead rabbit in its mouth, but it doesn't realize how to *eat the rabbit.* Many of these dogs are domesticated animals trying to survive in the wild."

But other dogs do manage to survive on the streets. Unless they're injured, Randy doesn't bother trying to rescue truly feral dogs that live in packs — oftentimes out of view of humans. "Some dogs are actually happier out here than they'd be in an apartment or a house," he said.

A good example was a large black dog Randy had been feeding for years, slowly earning its trust. The courtship was complete by my first visit to East St. Louis; the dog ran up to Randy's Jeep as we turned onto the block where he lived with his pack. The other dogs kept their distance — I could see their heads poking out from the vegetation of an overgrown field.

"How's my boy doing?" Randy said, kneeling on the ground to pet the dog, who wagged his tail in excitement.

"Have you thought about rescuing him?" I asked.

"All the time. But I don't think that's what he wants. He sort of runs the neighborhood. He has his harem."

Two days later, though, when the dog came hobbling up to us with a limp, Randy changed his mind. He lifted the black dog into his Jeep, but almost as soon as the dog was inside the car, it jumped back out.

"He obviously doesn't want to be rescued today," Randy said. "And I'm not going to force him. In his case, I think he actually knows what's best for him."

Most dogs don't jump voluntarily into Randy's Jeep. Most need some convincing.

After picking me up at the St. Louis RV Park on the first of my two days in town to see him this time, Randy drove us to North St. Louis, another area ravaged by poverty and violence. He'd been told that a stray

Pit Bull and her puppies were living in an overgrown yard behind a vacant house.

"Ready to save some doggie lives?" Randy asked me with a smile.

It was great to see him again. "I've been looking forward to this the whole trip!"

"How's the journey been? You meet any crazier dog-people than me?"

"You still get the gold," I assured him.

Though Randy seemed happy to see me, he hadn't yet recovered from what he'd witnessed a few days earlier in a vacant four-family building in North St. Louis.

"Benoit, it was the worst abuse I've ever seen," he said. "And I thought I'd seen everything."

Randy told me that five dogs had been tortured — they were burned, shot in the neck and spine, and strangled. When Randy had arrived at the scene, one dead Pit Bull was hanging out a window.

"When we catch whoever did this," Randy promised, "I want to be in the courtroom so I can make a big white-trash scene. It's going to be like the *Jerry Springer Show* up in there!"

A few minutes later, a detective called to update Randy on the investigation. The deaths had been widely reported in the news, and unlike so many other cruelty cases where Randy felt like Stray Rescue was the only organization that cared, this one had triggered a police investigation.

"What did the detective say?" I asked when Randy dropped his phone back in his lap.

"They're pulling out all the stops," he said. "They're going to try to get DNA evidence. There's going to be a reward. And they're talking to everyone to see who might know something. He told me they're going to ask the prostitutes in the area what they know, because prostitutes always know something. I'm not sure why everyone tells their secrets to prostitutes, but apparently they do."

Randy didn't have to drive over any bridges to get to North St. Louis on this day, so we made it to the Pit Bull and her pups without a panic attack. We couldn't find the dogs at first, but then Randy spotted the adult dog's nose poking out from inside a plastic igloo doghouse.

"Hey, little girl," Randy said, slowly making his way into the yard.

He'd told me earlier that when he rescues a dog he slows down his heart rate and breathing. "I try to go into a meditative state, which calms down the dog," he explained. "Everyone thinks I'm working some

493

kind of miracle, but it's not that. I'm just calm as a cucumber."

This particular stray — a beautiful white Pit Bull with cropped ears — was wary of Randy at first, but that was nothing his baby voice and a few bags of cold cuts couldn't fix. Before long, Randy had slipped an orange collar around her neck and was ordering me to grab a discarded mop bucket and pile the seven pups inside.

"We have to name the mom and the pups," he announced as we drove away with our adorable haul.

"Don't look at me — I'm terrible at dog names."

He lit a cigarette and turned up the volume on his car stereo, which was playing one of the songs he reserves for successful rescues — "Dancing on My Own" by Robyn.

"You write for a newspaper — let's name the dogs after papers," he suggested.

We settled on Time, News, Trib, Post, Dispatch, Star, Chronicle, and Enquirer. "I feel bad for whoever gets stuck with Enquirer," I said. "That's a terrible dog name."

As we drove back to Stray Rescue's headquarters to have the mom and pups checked out by a vet, we passed near an abandoned block where Randy and I had spent three days rescuing an emaciated Boxer mix in 2009. A friend of Randy's had alerted us to the dog, who was surviving on scraps from a gas station near a lumberyard and sleeping in a vacant house. We'd found her resting near an electrical pole in foot-high grass, a thick leather collar fastened tightly around her neck.

"Someone probably dumped her, or she got away," Randy had said as he'd parked the Jeep, grabbed a bag of hot dogs, and went about trying to bribe the animal with food. She was hungry, but she was also scared. Randy could only get within a few feet of her before she'd back away, quickly snatching a hot dog Randy left on the ground.

He eventually set up a cage trap, loading it with hot dogs and chunks of chicken. Then he called his mom. "Hey mom! I need a favor. I need something really good for the trap. Can you make me a roast? No, no, it doesn't have to be cooked all the way through. Dogs like it rare."

While we were on the subject of moms, Randy told me that dog-loving mothers can be the bane of a dog rescuer's existence. "You'll spend a week trying to get a dog, the dog finally goes in the trap you left overnight, and then the next morning some lady will get there before you do and say, 'Oh, poor dog! Someone trapped it!' And then she lets the dog go, thinking she's done some good in the world."

This particular dog turned out to be what Randy called "trap-savvy." She would tiptoe into the cage and crane her neck forward to grab a chunk of roasted chicken, careful not to step far enough inside to trigger the trap door.

While we sat in the Jeep and watched the dog outsmart us, Randy suggested we pass the time by playing one of his favorite games — Kill, Fuck, or Marry. "I'll pick three people, and you have to decide which one you would kill, which one you would fuck, and which one you would marry," he said.

"I know the rules," I assured him.

"I wasn't sure you would, being half-French and all."

"My American half is better developed," I said.

"Oh good. You know, you have to be able to have some fun when you're rescuing dogs," he went on. "It makes up for all the nights when I go home and cry." He opened the driver-side window and lit a cigarette. "Unfortunately, we don't have any mutual friends. So we'll have to use famous people."

He offered me three unenviable choices: Tom Arnold, Roseanne Barr, and Ed McMahon.

"Isn't Ed McMahon dead?" I asked.

"Oh, yeah. Let's say Ed McMahon a week before his death."

"That's in rather poor taste."

"That's the whole point of the game!"

I sighed. "I suppose I'd kill Ed McMahon, because he's near death anyway. I'd probably have sex with Tom Arnold, even though that visual disturbs me. And I guess I'd marry Roseanne."

"Oh, that sucks for you. Can you imagine being married to *Roseanne*?"

Thirty minutes and several games of Kill, Fuck, or Marry later, the dog had managed to clean most of the food out of the trap. Randy grudgingly decided to try to catch her with a net. Randy hates using his net because if he misses "the dog knows you're a bad person who's trying to trap her in a net. Any trust is gone."

Randy also happens to be a terrible net thrower. "What kind of dog rescuer are you?" I said as I watched him toss the net toward the dog, only to have it miss her by a good three feet. "That was the worst throw ever."

As Randy tried to figure out our next

move (his mother's roast wouldn't be ready until later that night), I confessed that I was hooked on dog rescuing.

"I get why you do this," I told him.

"Because I'm crazy?"

"No, because I've never felt so alive. I don't know how to explain it without sounding cheesy . . ."

"No, I get what you mean," he said, nodding.

"Usually when I'm doing something my mind is half there, half somewhere else. With you, out here, it's like the only thing that matters is the dogs. It's an adrenaline rush like I've never felt."

"Totally. The rest of my life can suck — hell, it usually does. I'm a mess. But when I'm out here, I forget about all that."

"I haven't checked my cell phone once today," I told him, genuinely astonished. "That *never* happens."

"Maybe you should move out here and rescue dogs with me," he said, smiling. "Can you believe I used to be a flight attendant?"

"Isn't that a problematic job for someone afraid of people and germs?"

"Yeah, trust me — you didn't want to be on my flight."

On our way back to Stray Rescue headquarters with the Pit Bull and her pups, Randy told me it was no surprise he ended up rescuing dogs for a living. As a kid, he'd lived under the tyranny of an abusive father — the only time his family got along was when Randy brought a stray dog home.

"I'm probably trying to fix my childhood with every dog I rescue," he said. "All my self-worth comes from dogs. Always has. These stray and feral dogs are my kids. They're what I live for. I think I'll probably be single forever, because all I really care about are these dogs."

Randy lives with eight rescues in his modest St. Louis home. "The few dogs that truly can't really be rehabilitated come to live with me," he told me. Randy is on the perpetual prowl for more foster and adoptive homes for the dogs he rescues, and he joked that he should start a website called StrayHarmony.com.

"It would be a site for lonely people to find the lonely stray dog of their dreams," he explained. "People could check a box that says 'I hate men,' and if the dog also hated men, we'd match them up. It's the perfect way to connect the freaks of the world with the freaky dogs of the world."

At a red light, an especially sad-looking shepherd mix shuffled across the pavement in front of us. "I haven't seen that dog before," Randy said. He jumped out of the car, kneeled down in the street next to an uncovered manhole, and fed the dog hot dogs while a car behind us honked.

The dog seemed desperate for this unexpected human kindness, even following Randy back to the Jeep. "Oh, honey, I can't take you with me today," he said. "I've got a full car already."

The dog got the picture, continuing on to wherever it was headed. We pulled away, and as Randy barreled down a highway toward Stray Rescue headquarters, I told him about nineteenth-century French writer Charles Baudelaire.

"You would have liked him," I said. "He loved street dogs and wrote the most beautiful poem about them."

I shuffled through my notebook, looking for where I'd noted Baudelaire's words. "Can I read the passage to you?"

"You better," he said.

I cleared my throat: "I sing the mangy dog, the pitiful, the homeless dog, the roving dog, the circus dog. . . . I sing the luckless dog who wanders alone through the winding ravines of huge cities, or the one who blinks up at some poor outcast of society with its soulful eyes, as much as to say, 'Take me with you, and out of our joint misery we will make a kind of happiness.' "

Randy didn't say anything at first. Then he wept.

Marc flew into St. Louis on my last night in town, and the next morning we hit the road on our way to western Indiana — and then Chicago.

While I drove us north through Indiana, Marc spent most of his time at the dinette studying for an important medical school exam. Every hour or so he'd make his way up to the front and massage my upper back, which, if it could speak, would have insisted I never again get behind the wheel of an RV.

We stopped at several dog parks on the way to Chicago. I'd discovered a handy iPhone app, DogGoes, which pointed us to nearby fenced-in parks where we could let the dogs loose. In Lafayette, Indiana, we paid $5 for a day pass to the Shamrock Dog Park, a membership-driven park accessible only with a keycard. I'd never had to pay to exercise my dogs, but Rezzy still couldn't be trusted off-leash in an open area — if she spotted a cat, rabbit, or squirrel, there was no telling how far she might chase it.

There was another reason Rezzy couldn't

be trusted. Cesar Millan had predicted that she might gain more confidence as she grew accustomed to her post-reservation life, but I never imagined that she would morph into a dog park bully. Rezzy had recently developed a dislike for some dogs larger than herself, especially if we were in an enclosed area. Twice in the previous few weeks I'd had to pull her off a dog. And she wasn't play-fighting — she'd hurled herself at the animals, teeth bared.

Rezzy's occasional aggression made me realize how much I'd been spoiled by Casey's California-style amiability. Though I could do without his humping, he's never been aggressive. If a dog comes after him, he turns his head away and defuses the situation. After a few weeks of Rezzy's unpredictability, I understood why a friend of mine had always seemed so tense — so miserable, even — while we walked our dogs together in a park. She never knew what her dog might do, which dog her dog might hurt.

Fortunately, Rezzy behaved herself at the Shamrock Dog Park. She spent some time sniffing an eleven-year-old Lab/Shar-Pei mix named Soapie, whose owner told us that the dog had been an accident: a Lab breeder and Shar-Pei breeder had lived next door to each other, and while the breeders never thought to introduce their dogs, a loose gate and the promise of sex brought the unlikely pair together.

"I guess we can call Soapie a *Labpei,*" Marc said.

The plan was to stay four days in the Chicago area — two at the home of a college friend's parents, and two at the pet-friendly Monaco Hotel in downtown Chicago. My friend's parents' house wasn't far from my alma mater, Northwestern, so on our first full day in Illinois I took Marc and the dogs on a tour of the campus. On a bright, clear late-spring afternoon, students read books or played Frisbee at the edge of Lake Michigan.

As nice as that walk was, Marc and I were going through a bit of a rough patch. We didn't acknowledge it then, but we both sensed it. I didn't feel as connected to him as I had on previous visits, and we were seemingly running out of things to talk about. Still, I wasn't going to panic over a couple days of distance. I still liked him.

On the day we were scheduled to head to Chicago and the Monaco Hotel, I joined my college friend's mom, a therapist named Sue, on a dog park excursion with Casey and Rezzy and her family's two Tibetan Ter-

503

riers. (Marc said he needed to stay at the house to study.) The dogs all had fun at the thirty-acre Independence Grove dog exercise area, and I took pictures of a white Ibizan Hound, a dog with the biggest ears I'd ever seen.

When we got back to Sue's, I walked upstairs to the guest room Marc and I were sharing. But Marc wasn't there. Neither was his suitcase.

On the bedside table, I spotted a note.

One of my favorite things about Casey is his predictability. After so many years together, I usually know what he's going to do well before he does.

That's how I knew he would lie down on the carpet next to me (without a sigh, since he'd just been on a long walk and was content) as I sat down on the bed to read the note and process what had just happened. Had Marc really just left? For the airport? Without saying good-bye? While I was at the *dog park*?

When I finally reached him an hour later on the phone, Marc explained that it was all happening too fast, that the pressure of medical school and our Chalet romance had gotten to him, that he wasn't sure we were meant for each other. He'd felt our distance the last few days, too, though he'd taken it as a sign that something was irrevocably broken.

"So instead of talking it out like an adult, you bolt?" I said. "What are you, a twelve-year-old?"

I was mad at him. But I was angrier with myself. How could I have liked someone who would act this way? What signs had I missed? Had the RV's heating or air-conditioning systems expelled crazy-making toxins? Until that moment, I'd prided myself on my taste in boyfriends. My previous relationships had all been with upstanding citizens — funny, smart, honest, and unlikely to scram without at least a conversation.

Of course, Marc wasn't really my *boyfriend,* and what we had wasn't quite a *relationship.* We'd had something, though. We'd opened up, let each other in. And I'd trusted him enough to involve him in a journey that wasn't supposed to include romantic intrigue. He'd been there — in person or on the phone — for some of the most important milestones, including the rescue of Rezzy. His fingerprints were all over the trip.

As I drove into Chicago that afternoon with the dogs, I worried that Marc's departure would taint my last week on the road. "How *dare* he try to ruin this for me," I moaned to my friend Dylan. "This trip wasn't about him, and now how he left is all I can think about."

I moped around Chicago for two days. I tried to motivate myself with an Al Pacino–esque pep talk from *Any Given Sunday (You will hold your head high today and fight through the pain and the doubt!),* but my self-pity lingered. I walked the dogs to the Cloud Gate public sculpture in Millennium Park, only to have a park employee tell me that dogs weren't allowed. I was getting sick of hearing that. I defiantly snapped pictures of Casey and Rezzy underneath Cloud Gate before agreeing to leave. I considered it my lone victory for the day.

I then took a preplanned, dog-friendly cruise around the Chicago River and Lake Michigan by Mercury Skyline Cruiseline, but my mood mirrored the gloomy, charcoal sky. When I was certain Marc wasn't going to magically reappear with flowers and an apology, I finally canceled the surprise dinner reservation I'd made for our last night there.

Thank God for Casey and Rezzy. Perhaps sensing that I needed to be handled delicately, Rezzy didn't attack any dogs over the next few days. If Casey humped anything, he did it when I wasn't looking. Back on the road in the RV, they mostly managed to fit together in their favorite spot.

I hurried east on Interstate 80, eager to get home. I'd planned to stop in Elkhart, Indiana, at the northern tip of the state and just off the interstate. Elkhart is considered the "RV Capital of the World," and part of me wanted to visit. But the part of me that wanted the trip to be over won out.

Before too long, though, the dogs slowed me down. They needed to pee, needed to run, needed a respite from my human drama. They dragged me out of the RV and, slowly, out of my despondency.

At a truck stop on the Ohio Turnpike, Casey and Rezzy led me over to a rescue mutt (Bella) and her truck-driving human (Dave).

"Where are you and the dogs headed?" Dave wanted to know.

"Massachusetts," I told him. "Can't get back soon enough. I've been on the road for almost four months, and I'm having a bad few days."

"RV trouble?" he wondered.

I shook my head. "Boy trouble."

"Oh," he said.

I asked if I could pet Bella, who was

standing triumphantly in the truck's front cab, her head and chest visible out the passenger-side door.

"I wouldn't recommend it," he told me. "I'm one of the few people she likes."

Dave recalled that when he'd decided to start driving a truck for a living, he went looking for a dog he could take along for the ride. At the Humane Society in Payson, Arizona, he'd met Bella, who had been at the shelter for months because of her aggression toward humans. Bella, though, shocked the shelter staff when she took an immediate liking to Dave during his visit. (Unlike actor George Clooney, Dave didn't need to resort to trickery to impress the dog. Clooney once rubbed turkey meatballs on his shoes to ensure that a shelter dog he coveted would show reciprocal interest.)

"I guess there really is a dog for everybody," I said.

"Yeah, though I'm not sure what it says about her taste — there are better guys out there than me," Dave told me. He shrugged his shoulders. "Still, I'm not complaining."

"What's the greatest thing about having her with you in the truck?" I asked him a couple minutes later, after Bella jumped out of the cab and came to stand next to him.

He didn't hesitate. "We keep each other company."

We keep each other company.

Is the dog-human bond as simple as that? Over the course of my fifteen weeks on the road, I'd spent more than a few nights in barren RV parks reading about complicated anthrozoological and psychological theories of pet ownership. Some people believe we share our lives with dogs because they're handy substitutes for human interaction in an increasingly disconnected world. Others think we live with dogs because they keep us active and healthy, or because they teach empathy and responsibility to children.

There are also less generous theories, including that we live with pets because of a subconscious need to dominate nature, or because "of a misfiring of our parental instincts," as Hal Herzog put it. Herzog offered another possible explanation: "From the meme's-eye view, pet-keeping is a mental virus spread by imitation," he writes. He concedes that while "this idea seems far-fetched . . . the evidence for this perverse hypothesis is surprisingly strong."

Herzog notes that living with a dog in Japan became popular only after World War II, when "the Japanese began to emulate aspects of American culture." Or that in Sri Lanka, the higher power you choose to believe in predicts whether you'll share your life with a dog. "The fact that a Sri Lankan Buddhist is twenty times more likely than a Muslim to own a dog," Herzog writes, "suggests that Islam inoculates believers from infection by puppy-love memes, while Buddhism makes people more susceptible."

But I like the *we keep each other company* theory of pet ownership best. Whatever the merits of the ownership-as-virus in the abstract, I'd encountered no evidence of it on my trip. The dog lovers I met didn't live with dogs out of some rote, unthinking social obligation; they lived with dogs because they couldn't imagine life without them. Even many of those who relied on canines for a specific skill (herding, police work) couldn't help letting their dogs finagle their way into the master bedroom — and their heart.

When my friends called me during my journey to gush about their dogs ("Little Petey wants to be in your book!"), they spoke often about the ease of their dog's company. The dog could be counted on — to show up, to make them laugh, to comfort them — in a way that many of the humans in their lives couldn't, or wouldn't. In many cases, though, their close human-animal bond came only after some initial conflict or misunderstanding. There had often been some drama to overcome.

One of my favorite examples of inauspicious beginnings came courtesy of a friend, Joe, who runs a landscape nursery near Boston. He'd called me when I was in Truth or Consequences to see how I was holding up. I told him that I would never forget when he took me aside and recounted the first few days he'd spent with his Otterhound mix, Winona.

The dog had quite a backstory: she and her littermates had been tossed from a moving vehicle in Tennessee and ended up at a local no-kill shelter. While Winona's siblings were quickly adopted, she spent two years at the shelter. Nobody wanted her — until Joe did.

He arranged to have her transported by a trucking company that brings Southern shelter dogs to their new lives in the Northeast. At a truck stop in Connecticut, Joe stood alongside a dozen other people waiting for their dogs. One by one, the rescues were led out on a leash to meet their human families.

"Finally, a handler came out with the scruffiest, most timid, and wildly shaking animal I'd ever seen," Joe recalled. "She looked something like a cross between a pig and an opossum. That was Winona."

The dog froze when the handler gave Joe the leash. As Joe kneeled down next to her, Winona wormed her way out of her collar and ran straight for the highway. Fortunately, she took a pit stop behind the wheel of the transport truck, giving the handler time to catch the trembling dog, secure her with a new collar, and hand her back to Joe.

He drove Winona to a nearby hotel, where he hoped to bond with her before bringing her to his home in Massachusetts. But on the way, Winona pooped all over the car's backseat — and herself.

In their hotel room, Joe gave the terrified dog a bath, after which Winona wedged herself into a tiny space near the nightstand and refused to budge. Hoping food might help his dog trust him, Joe handfed her a burger from the hotel's café. A few minutes later, Joe coaxed Winona out of their room for a walk on the lawn, but when they returned she refused to follow him inside.

Joe picked up his quivering dog and placed her on one of the room's two queen beds. "Then poop basically started exploding out of her again," Joe said. "She ran frantically around the room, feces flying everywhere."

There were coin laundry machines across the hall, so Joe spent the next few hours washing sheets and blankets. Then he scrubbed the carpets and went to sleep.

The next morning, Joe tried to sweet-talk Winona into his car for the drive to Massachusetts. He picked her up with "all the deliberate tenderness I could muster," but in the car Winona was frantic and started pooping again.

"So there we were, with shit everywhere and me cleaning and trying to calm her down," Joe said. "I felt so bad for Winona. I'd never seen a dog so scared, and I was worried that I was hurting her as I tried to clean her. Finally I looked at her and I said, 'I am never going to leave you. I will clean you up as long as it takes. I will *never* abandon you.' Then I burst into tears. And you know what that little rat-pig-possum did? She stuck out her long tongue and licked all my tears away. That was the end of the pooping. Benoit, it was like she understood me. I'm *convinced* she understood me. Now, every time she sees me throughout a day, her tail wags in circles so hard, she literally looks like a helicopter about to take off!"

■ ■ ■ ■

For my last night on the road, I parked the Chalet at a KOA southeast of Pittsburgh. I was grateful to be feeling sentimental; it meant I wasn't thinking about Marc.

"Today is my last full day in the El Monte RV that has been my home these past four months," I wrote on the trip's Facebook page. "Things I will miss about my 25-foot ride: Playing my harmonica while lounging in the shockingly comfortable bed . . . Easy access to midnight snacks . . . Meandering along in the right lane to Florence and the Machine . . . The freedom to drive my home wherever I want to go."

Then I added what I wouldn't miss about the RV life: "Spending between $150 and $190 to fill up my home's gas tank every 400 miles. Small country roads with over-sized trucks barreling toward me. Wind, which throws the RV around and makes me seem drunk. Emptying out the 'black water' tank. Waking up in a panic, unsure what state I'm in."

I took Casey and Rezzy for a walk around the park's small lake, where a mother and her young son fished with the tiniest poles I'd ever seen. "Catch anything?" I asked, though I doubted they could reel in a goldfish with their equipment.

I let Casey off the leash and watched him saunter to the water's edge. He sniffed some grass, then steadied himself to pee on it. (Unless he's urinating against a tree or a fire hydrant, Casey doesn't lift his leg.)

"You're a good boy, Casey," I said, prompting him to look back at me with that slightly confused look he gets when I compliment him for doing nothing.

My session with Dr. Gold felt like it had happened years ago, in a parallel universe. He'd expressed doubt that this road trip would fundamentally alter the nature of my relationship with Casey, but it had.

What had caused the shift? I wasn't exactly sure. I suspect that I'd learned something from the hundreds of dog people I had had the fortune to meet during my long American walkabout. (Or *Chalet*-about, as it were.) Sure, some worried whether their dogs were happy. Others confessed that they didn't always feel like they deserved their dog's attention, respect, or blind adoration. But most simply appreciated their dogs. Their dogs weren't the cause of their worries; they were a respite from them.

Watching Casey interact with so many different people and dogs helped me realize

515

just what a trouper he is. He's easy to be around, easy to entertain, and easy to travel with. (It helped that he finally learned to tolerate the RV.) He loves being outside, loves being athletic. He's friendly and kind, if sometimes distant and a little neurotic. In all those ways, he's a lot like me.

Somewhere around Texas, I stopped worrying about whether Casey was the right dog for me and started appreciating him for the dog he is. (I also came to appreciate simply having an active, healthy dog. Andy, the Pembroke Welsh Corgi I'd spent hours searching for in Connecticut, had never been found.)

Still, I had to worry about *something*. So I turned my attention to Rezzy, who had spied a squirrel in a nearby tree and was straining at the leash and whining ever so softly. She wanted desperately to reach the squirrel, to live out her calling as a chaser of animals big and small.

I was more convinced than ever that Rezzy was part, or mostly, Border Collie. For one thing, she looked like a Border Collie. For another, she loved running after things — and herding Casey. Had I rescued her before my stop at Rob and Bruce's cattle ranch in Colorado, I would have let her loose on the cows. I bet she would have known exactly what to do.

Part of me worried that I'd made the wrong decision rescuing Rezzy. On the reservation, she could chase anything she liked. In fact, she'd been after something in a field the moment I'd pulled into the Spirit gas station. Should I have let her stay out there with her pack? When I'd called Garrett to confess my Rezzy worries during a long drive through Montana, he laughed.

"You idiot, she would have *died* if we hadn't taken her," he reminded me. "Remember her uterine infection?"

"Oh, yeah." How had I forgotten about that?

But if Rezzy was in fact a Border Collie, was she meant to live in a city with me? Sure, I would take her to parks and let her run free. But wouldn't she be happier on a farm somewhere? Wouldn't she prefer to live somewhere else, with someone else?

I sat on the grass by the lake. The squirrel was long gone, so Rezzy shuffled over to me and nuzzled her snout into my chest.

"Are you sure you want to live with me?" I asked her.

I brushed my teeth, closed the RV's shades, and powered down my laptop. There were a few dirty dishes in the sink, but I figured I could get to those in the morning.

No reason to start being a neat freak on the last night of my trip.

I encouraged the dogs to join me on my bed. They jumped up, tails wagging. When Casey realized we weren't going to play a game, he sighed and plopped himself down at my feet. Rezzy, meanwhile, edged toward my pillow. She wanted to be close to me.

I don't remember what I dreamt about that night. But when I woke up the next morning to the sound of kids playing by the lake, the dogs of my dreams were still there.

They'd barely moved an inch.

ACKNOWLEDGMENTS

Though I traveled around the country with a dog (and then two dogs), it took a village of humans to keep me alive and on schedule — and to help me complete the book when I returned home.

The biggest thanks go to Egan Millard and Jenny Kutner, two dog-loving young writers/journalists without whose help I'd surely still be holed up in my messy home office, trying to finish the manuscript while Casey sighs, exasperated, at my feet.

A number of other folks helped in important ways, whether it was researching a certain topic or pointing me toward an article or book I shouldn't miss. Special thanks to Adam Polaski, Nick Lehr, Joe Caputo, Brandon Ambrosino, Myles Tanzer, Taylor Grow, Hayden Wright, Seth Putnam, and August Thompson.

A handful of friends read the manuscript (in part or in full) and offered valuable suggestions. I am grateful to Bob Smith, Kevin Sessums, Jameson Fitzpatrick, Rachel Kadish, Matt Heller, Henry Paul Belanger, Zach Wichter, Neil Savage, Sam Bickett, Bryan Lowder, Darcia Murrish, Zac Bissonnette, Ian Shin, Kate Beutner, Duncan Roy, Dietrich Warner, and Dave Ford. And a very big thank-you to my good friend James Doty, a man who dislikes most dogs but who nonetheless edited my manuscript with aplomb, only occasionally complaining that he'd rather be watching the Tennis Channel.

My cross-country journey would have been significantly less fun without brief visits from Garrett Beltis and Sam Reid, who brought me much-needed company and levity. (Garrett gets bonus points for taking amazing pictures and videos of our time in the Southwest.)

A number of dog people generously helped me plan my journey, suggesting places, people, and dogs I shouldn't miss — these helpful humans include Amy and Rod Burkert, Bob Aniello, James Serpell, Alice Kaplan, Stephanie Hunnicutt, Buck Johnston, Ryan Clinton, Kelly Hardie, Dr. Joel Gavriele-Gold, Julie Hecht, Christopher Castellani, Leslie Smith, Lucy Maloney, Joseph Dwyer Jr., David Yaskulka, Kelly Ausland, and Lynne Ouchida.

Though my El Monte motorhome was comfortable, I rarely turned down an offer to sleep in a bedroom without wheels. I'm grateful for the generosity of Riverbend Hot Springs in New Mexico, the Flint Hill Public House & Country Inn in Virginia, the Sorrento Hotel in Seattle, and the Kimpton Monaco hotels in Portland and Chicago. A handful of friends also offered me food and shelter. Special thanks to Darcia Murrish, Mike Guadagno, Rob Hoiting, and Brant Granger and Neil Ingalls.

I spent time with many amazing people on my journey, but I want to thank two in particular — Randy Grim and Beth Joy Knutsen — for their support and friendship. I also had the privilege of meeting the staff and volunteers at many great rescue and dog organizations, not all of whom I was able to write about in the book. Thanks to the folks at Austin Pets Alive!, San Antonio Pets Alive!, Animal Rescue New Orleans, Best Friends Animal Sanctuary, the Dakin Pioneer Valley Humane Society, Gabriel's Angels, freekibble.com, and dogtime.com.

I'm grateful to photographers Amanda Jones and Brad DeCecco, who contributed photographs for the book. Amanda took the wonderful picture of Casey on the cover.

Though I failed in my efforts to get a fuel sponsor for my trip, other great companies stepped up with generous support: El Monte offered me a discount on a motorhome, Halo Purely for Pets sent me enough dog food to feed Casey (and then Rezzy), and Kampgrounds of America gave me free passes to many of its locations.

A writer needs time to write, and I was fortunate to receive support from The Mac-Dowell Colony, The College of Wooster, and Emerson College (where I teach in the department of Writing, Literature & Publishing).

As usual, the team at Simon & Schuster — including editor Bob Bender and publisher Jonathan Karp — made the book writing and editing process a pleasant one (or as pleasant as book writing and editing can be). Thanks also to my longtime agent, Todd Shuster, who never wavers in his support.

Finally, and most importantly, I want to thank Casey and Rezzy. The dogs of my dreams, to be sure.

NOTES

Prologue

"in all likelihood dogs do not make": Caroline Knapp, *Pack of Two: The Intricate Bond Between People and Dogs* (New York: Delta, 1998), p. 108.

In an article he wrote trumpeting: Joel Gavriele-Gold, "Dogs in a Psychologist's Office," *AKC Gazette,* July 1993. Article available on Gavriele-Gold's website, http://www.drjoelgold.com/gazette.html.

"the unconscious act of putting": Joel Gavriele-Gold, *When Pets Come Between Partners* (New York: Howell Book House, 2000), p. 5.

"someone might fear": Ibid.

Chapter 1

"There is a hope that a dog injects": John Zeaman, *Dog Walks Man: A Six-Legged*

Odyssey (Connecticut: Lyons Press, 2010), p. ix.

"always want to take": Roger Grenier (translated by Alice Kaplan), *The Difficulty of Being a Dog* (Chicago: University of Chicago Press, 2000), p. 25.

"She was the only one in our room": Associated Press, Jan. 5, 2012.

a study by psychologist: A. Horowitz, "Disambiguating the 'Guilty Look': Salient Prompts to a Familiar Dog Behavior," *Behavioural Processes* 81, no. 3 (2009): 447–52.

You gonna eat that?: Karen Shepard, "Birch," in Amy Hempel and Jim Shepard, eds., *Unleashed: Poems by Writers' Dogs* (New York: Crown, 2007), p. 30.

Judges in animal disputes tend to rule in favor of women: Stanley Coren, *The Modern Dog: A Joyful Exploration of How We Live with Dogs Today* (New York: Free Press, 2008), p. 184.

"a dog is the legal equivalent": Ibid., p. 182.

a "child substitute": Orange County Superior Court Judge John Wooley said this in 1983, as reported by the Associated Press.

In one unusual custody dispute: Coren, *The Modern Dog,* p. 184.

Massachusetts law that forbids frightening a

pigeon: Section 132 of Chapter 266 of the General Laws of Massachusetts.

"a kind of victory over the anonymity": Mary Battiata, "Lassie Go Home," *Washington Post,* May 30, 1999.

"And I know where they go, these women": Amy Hempel, "In the Animal Shelter," *The Collected Stories of Amy Hempel* (New York: Scribner, 2006), p. 157.

the demographic group least likely: Coren, *The Modern Dog,* p. 232.

"the iconic American road book": Bill Steigerwald, "Sorry, Charley," *Reason,* April 2011.

"He just sat in his camper": Charles McGrath, "A Reality Check for Steinbeck and Charley," *New York Times,* April 4, 2011.

believes Steinbeck "made up most of the book": Ibid.

"Charley is a born diplomat": John Steinbeck, *Travels with Charley* (New York: Penguin, 2012), pp. 5–6.

"would rather travel about than anything he can imagine": Ibid., p. 6.

"A journey is a person in itself": Ibid., p. 1.

The New York Times *noted in 1990:* Andy Grundberg, "Coping; With Pet Portraiture," *New York Times,* June 30, 1990.

"it is a difficult matter to get": "In Amateur

Photography's Field," *New York Times,* March 15, 1891.

in one study, nearly 70 percent of four-year-olds: K. Meints et al., "How to Prevent Dog Bite Injuries? Children Misinterpret Dogs' Facial Expressions," *Injury Prevention* 16 (2010): A68.

"each speaking our own native 'language' ": Patricia McConnell, *The Other End of the Leash: Why We Do What We Do Around Dogs* (New York: Ballantine, 2002), p. xviii.

"The tendency to want to hug something": Ibid., p. xxi.

"on the way to somewhere else, somewhere better": Rick Moody, *State by State: A Panoramic Portrait of America* (New York: Ecco, 2009), p. 67.

"Of all the beautiful towns": Mark Twain, in a letter to *The Alta California,* Sept. 6, 1868.

Pynchon later recounted in an article for Connecticut Quarterly: W. H. C. Pynchon, *Connecticut Quarterly,* April/June 1898.

Yale professor Blair Kauffman told the Yale Daily News: Nikita Lalwani, "YLS Dog Rentals to Continue," *Yale Daily News,* April 21, 2001.

who rank seventy-seventh out of seventy-nine in dog intelligence: Stanley Coren, *The*

Intelligence of Dogs: A Guide to the Thoughts, Emotions, and Inner Lives of Our Canine Companions (New York: Atria, 2006).

If a 2011 morning show segment: "Dognapping on the Rise: How to Protect Your Dog," *Good Morning America,* Aug. 18, 2001.

According to a 2012 study by the ASPCA: E. Weiss et al., "Frequency of Lost Dogs and Cats in the United States and the Methods Used to Locate Them," *Animals* 2, no. 2 (2012): 301–15.

Chapter 2

39 percent of French dogs: L. Colliard et al., "Risk Factors for Obesity in Dogs in France," *The Journal of Nutrition* 136, no. 7 (July 2006): 1951S–54S.

In a 1975 letter to The New York Times: Carol Hillman, Letter to the Editor, *New York Times,* May 8, 1975.

"Like the Jews of Nazi Germany": Gini Kopecky, "To Scoop or Not to Scoop," *New York Times Magazine,* Aug. 20, 1972.

"Pets should be disallowed by law": Fran Lebowitz, *Social Studies* (New York: Pocket, 1982), p. 54.

"both the grass shortage and the poop sur-

plus": Michael Brandow, *New York's Poop Scoop Law: Dogs, the Dirt, and Due Process* (West Lafayette, IN: Purdue University Press, 2008), p. 260.

"were not about to hand over parcels": Ibid., p. 72.

"Critics imagined horrible scenes of loose predators": Ibid., p. 261.

In a French study, women were significantly: N. Gueguen et al., "Domestic Dogs as Facilitators in Social Interaction: An Evaluation of Helping and Courtship Behaviors," *Anthrozoos* 21 (2008): 339–49.

"The results suggest a human": A. Fridlund et al., "Approaches to Goldie: A Field Study of Human Approach Responses to Canine Juvenescence," *Anthrozoos* 12 (1997); 11(2):95–100.

Researchers have found that children: Coren, *The Modern Dog,* p. 147.

"made the children more cooperative": Daniel Goleman, "Children and Their Pets: Unexpected Psychological Benefits," *New York Times,* Jan. 11, 1990.

In a 2003 New York Times *piece:* Caroline Campion, "Straining at the Leash," *New York Times,* March 16, 2003.

"Homeless people were moved": Ibid.

"Admirers of the Xolo concede": Guy Trebay,

"Who Are You Calling Ugly?," *New York Times,* Feb. 11, 2011.

"the nose with paws": McConnell, *The Other End of the Leash,* p. 67.

"fit inside the sleeves of a Chinese nobleman's robes": Josh Dean, *Show Dog: The Charmed Life and Trying Times of a Near-Perfect Purebred* (New York: HarperCollins, 2012).

"producing unhealthy freaks": Raymond and Lorna Coppinger, *Dogs: A New Understanding of Canine Origin, Behavior, and Evolution* (Chicago: University of Chicago Press, 2001), p. 248.

I'd written a cover story: Benoit Denizet-Lewis, "Can the Bulldog Be Saved?," *New York Times Magazine,* Nov. 22, 2011.

"poster child for breeding gone awry": Lee Shearer, "Standards for Breed Altered," *Augusta Chronicle* (ME), Jan. 16, 2009.

"Many would question": Dr. Nicola Rooney and Dr. David Sargan, "Pedigree Dog Breeding in the U.K.: A Major Welfare Concern," an independent report commissioned by the RSPCA.

"facilitate the attribution": James Serpell, *Thinking with Animals,* "People in Disguise: Anthropomorphism and the Human-Pet Relationship" (New York: Columbia Uni-

versity Press), 2005.

"if bulldogs were the product": Ibid.

"scarcely capable of any education": William Youatt, *The Dog* (Philadelphia: Blanchard and Lea, 1855), p. 150.

"Nobody who is anybody": Mark Easton, *Britain Etc: The Way We Live and How We Got Here* (London: Simon & Schuster UK, 2012).

Chapter 3

A team of psychologists: S. Gosling et al., "Personalities of Self-Identified 'Dog People' and 'Cat People,' " *Anthrozoos* 23, no. 3 (2010): 213–22.

"In this way": Alexandra Horowitz, *Inside of a Dog: What Dogs See, Smell, and Know* (New York: Simon & Schuster, 2010), p. 84.

"I finally got a friend": Huma Khan, "White House Unveils Excited First Pup," ABC News, April 14, 2009.

George Washington, who is credited: Roy Rowan and Brooke Janis, *First Dogs: American Presidents and Their Best Friends* (New York: Algonquin, 2009), pp. 15–16.

When a Fox Terrier wandered: Ibid.

When John Adams moved into the White House: Amelia Glynn, "Celebrating Our

Founding Fathers' Furry Friends," SF-Gate.com, June 30, 2010.

I participate in all your hostility to dogs: Rowan and Janis, *First Dogs,* p. 18.

While an AKC online poll revealed: Stacy St. Clair, "Obama Promises Daughters a Puppy in the White House," *Chicago Tribune,* Nov. 5, 2008.

"This will fuel the breeding industry": Nick Greene, "Obamas Criticized for Breaking Promise of Adopting Dog from Shelter," London, England, *The Telegraph,* April 14, 2009.

strapping his Irish Setter to the roof: Ana Marie Cox, "Romney's Cruel Canine Vacation," *Time,* June 27, 2007.

Comedian Dennis Miller once joked: "Dennis Miller on Criticism of Obama, PETA vs. the Circus, and Spring Break in Mexico," FoxNews.com, March 26, 2009.

Some of the biggest names in clothing: "Mango Bans Exotic-Animal Skins Following PETA Meeting," Peta.org, November 17, 2011.

and fast food: Michael Yaziji and Jonathan Doh, "Case Illustration: PETA and KFC," *NGOs and Corporations* (Cambridge, England: Cambridge University Press, 2009), pp. 112–14.

"This is not a democratic organization": I Am

an Animal: The Story of Ingrid Newkirk and PETA, HBO, 2007.

In the New Yorker *profile:* Michael Specter, "The Extremist," *The New Yorker,* April 14, 2003.

"romantic and timeless . . . interspecies contract": Lars Eighner, *Travels with Lizbeth: Three Years on the Road and on the Streets* (New York: Fawcett Columbine, 1993), p. xii.

who had "the now old-fashioned idea": John Homans, *What's a Dog For? The Surprising History, Science, Philosophy, and Politics of Man's Best Friend* (New York: Penguin, 2012), p. 48.

Let's Have a Dog Party!: Ingrid Newkirk (Avon, MA: Adams Media, 2007).

In 2011, PETA euthanized 95 percent: Janice Lloyd, "Peta Says 'Exploiters' Raise Euthanasia Issue," *USA Today,* March 1, 2012.

"talk to deaf dogs": McConnell, *The Other End of the Leash,* p. 40.

"Yet he looked like Max": Coren, *The Modern Dog,* p. 42.

dog trainer Brian Kilcommons likes to joke: George Greenfield, *The Complete Book of Pet Names: An ASPCA Book* (Kansas City: Andrews & McMeel, 1997), p. x.

Pet photographer Walter Chandoha: Ibid.

"a name for a pet is, or at least should be": Ibid., p. ix.

"Whimsy and humor can generally play": Ibid.

In a later paper on the subject: Amanda Leonard, "The Plight of 'Big Black Dogs' in American Animal Shelters: Color-Based Canine Discrimination," *Kroeber Anthropological Society* 99, no. 1 (2011): 168–83.

"Who understands the menu of stink better?": Henri Michaux quoted in *The Difficulty of Being a Dog,* Roger Grenier (Chicago: The University of Chicago Press, 2000), p. 12.

would let "the dog poop on purpose": Justin Jouvenal, "Fairfax Neighbors Head to Court over Unscooped Dog Poop," *Washington Post,* Oct. 24, 2011.

"not once during these three days": Ibid.

"Will it offend you if I use the word 'poop'?": Julie Carey, "Fairfax County Woman Cleared in Dog Poop Trial," NBCwashington.com, Oct. 25, 2011.

"Do dogs really use the bathroom?": Ibid.

in 2008 the sisters had filed an unsuccessful complaint: Jouvenal, "Fairfax Neighbors Head to Court over Unscooped Dog Poop."

"fecal matter": R. Bowers et al., "Sources of Bacteria in Outdoor Air Across Cities in

the Midwestern United States," *Applied and Environmental Microbiology* 77, no. 18 (Sept. 2011): 6350–58.

"I don't always pick up": Zeaman, *Dog Walks Man,* pp. 32–33.

the highest prize at the San Francisco Chronicle: Mary Ann Dancisin, "Virginia Wineries: Best in Class, and Double Gold, in SF Competition," *Virginia Wine Guide Online.*

There's a great YouTube video: "Friendly Dog Meeting Curious Cows Makes Internet Hit," London *Telegraph,* Jan. 22, 2012.

A blind man, Bill Irwin: Bill Irwin, *Blind Courage* (Waco: WRS Publishing, 1996).

Chapter 4

According to a survey of American dog owners: "Thundershirt Study Reveals Prevalence of Anxiety Among Dogs," Thundershirt.com.

"the ninja warriors of dogdom": Richard Woodward, "Great Plott!," *Slate,* Feb. 12, 2008.

"no other animal has been portrayed so ubiquitously as the Bad Guy": Brian Hare and Vanessa Woods, *The Genius of Dogs: How Dogs Are Smarter than You Think* (New York: Dutton, 2013), p. 17.

while one recent study suggests a much earlier start date: K. Wayne et al. "Complete Mitochondrial Genomes of Ancient Canids Suggest a European Origin of Domestic Dogs," *Science* 342, no. 6160 (2013): 871–74.

a fitting touch considering Greene died of sunstroke: "Then and Now: A Last Walk in the Sun," Warwickonline.com, Nov. 19, 2010.

"What do you mean Patrick is dead?": John Berendt, *Midnight in the Garden of Good and Evil: A Savannah Story* (New York: Vintage, 1999).

but that a "chimney effect": "Where in the World Is Stuckie?," walb.com, Feb. 21, 2003.

Chapter 5

"With her makeup off": Joe Friedman, "Why I Hate Florida," JoeFriedman.hubpages.com.

Americans spent $53 billion on pet products: American Pet Products Association.

The pet industry has proven: Carol Tice, "Why One Recession-Proof Industry Just Keeps Growing," *Forbes,* Oct. 30, 2012.

including a New York Times *piece:* Andrew Martin, "For the Dogs Has a Whole New

Meaning," *New York Times,* June 4, 2011.

"the desire to treat pets as humans": Michael Shaffer, *One Nation Under Dog: Adventures in the New World of Prozac-Popping Puppies, Dog-Park Politics, and Organic Pet Food* (New York: Henry Holt, 2009), pp. 23–24.

"Nearly any trend in human consumerism": Ibid.

"The U.S. pet set gets not only": "The Great American Animal Farm," *Time,* Dec. 23, 1974.

"attributing human qualities": Alexander Hammer, "Affluence Is Fueling Pet Industry Growth," *New York Times,* Sept. 21, 1969.

A 2008 University of Chicago study: N. Epley et al., "When We Need a Human: Motivational Determinants of Anthropomorphism," *Social Cognition* 26, no. 2 (2008): 143–55.

Studies also tell us: M. McPherson et al., "Social Isolation in America: Changes in Core Discussion Networks over Two Decades," *American Sociological Review* 71, no. 3 (June 2006): 353–75.

"the dog has been granted temporary personhood": James Serpell, *The Domestic Dog* (Cambridge, England: Cambridge University Press, 1996), p. 254.

"Our expectations are really going up": James Vlahos, "Pill-Popping Pets," *New York Times Magazine,* July 13, 2008.

4.5 million people in this country are bitten: J. Sacks et al., "Dog Bites: How Big a Problem?," *Injury Prevention* 2 (1996): 52–54.

"if you want to cause a commotion": Coren, *The Modern Dog,* p. 116.

flat-out equate dogs to people: Gregory Berns, "Dogs Are People, Too," *New York Times,* Oct. 5, 2013.

Erica devised an experiment: E. Feuerbacher and C. Wynne, "Relative Efficacy of Human Social Interaction and Food as Reinforcers for Domestic Dogs and Hand-Reared Wolves," *Journal of the Experimental Analysis of Behavior* 98, no. 1 (July 2012): 105–29.

the author of a study about dogs and scent: N. Hall et al., "Training Domestic Dogs (*Canis lupus familiaris*) on a Novel Discrete Trials Odor-Detection Task," *Learning and Motivation* 44, no. 4 (Nov. 2013): 218–28.

a German Shepherd has some 225 million: Stanley Coren, "Do Some Dog Breeds Have Better Noses and Scent Discrimination than Others?," *Psychology Today,* Jan. 15, 2011.

"people may behave like animals": "Dogs, wolves, and the meaning of 'dogness,' " published on the University of Florida website, June 15, 2010.

"Do Dogs (Canis familiaris) Seek Help in an Emergency?": K. Macpherson and W. Roberts, "Do Dogs (*Canis familiaris*) Seek Help in an Emergency?," *Journal of Comparative Psychology* 120, no. 2 (May 2006): 113–19.

"it ran over and jumped in the person's lap": Alex Boese, *Elephants on Acid: And Other Bizarre Experiments* (New York: Harvest, 2007).

"arguably the most successful mammal": Hare and Woods, *The Genius of Dogs*.

"that a ball cannot pass through a solid object": Ibid., p. 155.

The Genius of Dogs *includes a handy chart*: Ibid., p. 235.

"where no matter what they could see": Ibid., p. 151.

have your dog microchipped instead: Ibid., p. 154.

"will inspect different hiding locations": Ibid., p. 161.

Studies designed specifically: Ibid., p. 161.

"it's become a cliché that we love dogs": McConnell, *The Other End of the Leash*, p. 7.

"that our pets are programmed": Hal Herzog, *Some We Love, Some We Hate, Some We Eat* (New York: Harper, 2010), p. 79.

were believed to be "indefatigable mourners": Susan Orlean, *Rin Tin Tin: The Life and the Legend* (New York: Simon & Schuster, 2012), p. 116.

"A dog's life is a short one": Ibid.

they "inflict the suffering of loss upon us": Grenier, *The Difficulty of Being a Dog,* p. 9.

Chapter 6

"the perfect teacher": Pema Chodron, *When Things Fall Apart* (Boston: Shambhala, 2005), p. 14.

"one of those alarming West Texas sunsets": Michael Hall, "The Buzz About Marfa Is Just Crazy," *Texas Monthly,* Sept. 2004.

"Simply entering virtually assures an animal's doom": Jean Martin, "Death by Pound," *San Antonio Express News,* Nov. 21, 2004.

"don't like restrictions on their dogs": Herzog, *Some We Love, Some We Hate, Some We Eat,* p. 126.

"health, behavior, and nutrition": Schaffer, *One Nation Under Dog,* p. 224.

"people giving up their pets": Ibid.

"animals, which do nothing useless": Grenier, *The Difficulty of Being a Dog,* p. 3.

"Bed Bugs: Old Problem Resurfaces in RV Parks": Leanne Phillips, "Bedbugs: Old Problem Resurfaces in RV Parks," rvbusiness.com, Feb. 28, 2011.

"shouldn't disregard the possibility": Jeremy Ecker, "Can Dogs Carry Bed Bugs," TheBedBugInspectors.com, Sept. 11, 2012.

"The sky was as full of motion and change": Willa Cather, *Death Comes for the Archbishop* (New York: Vintage, 1990), p. 231.

"many working-class blacks are easily intimidated by strange dogs": Elijah Anderson, *Streetwise: Race, Class, and Change in an Urban Community* (Chicago: University of Chicago Press, 1992).

"African Americans in particular may endorse": K. Chapman et al., "Fear Factors: Cross Validation of Specific Phobia Domains in a Community-Based Sample of African American Adults," *Journal of Anxiety Disorders* 25 (2011): 539–44.

Chapter 7

to the hundreds of thousands: Jeri Clausin, "Navajo Nation's Dogs Roam Unchecked," *Huffington Post,* Aug. 16, 2011.

there isn't a single veterinarian: Julie Hauserman, "Lakota Community Offers Help for

Reservation Dogs," Humane Society of the United States, Jan. 5, 2011.

The Choctaw nation: John Upton Terrell, *American Indian Almanac* (Cleveland: World, 1971).

The Iroquois tribes used to: Harold Blau, "The Iroquois White Dog Sacrifice: Its Evolution and Symbolism," *Ethnohistory* 11, no. 2 (Spring 1964).

According to a belief common among Plains Indians: Harold E. Driver, *Indians of North America* (Chicago: University of Chicago Press, 1969), p. 34.

Unlike tribes who partook of dog meat: Terrell, *American Indian Almanac,* p. 289.

Several Plains tribes, including the Shoshone: Robert H. Lowie, *Indians of the Plains* (New York: McGraw-Hill, 1954), p. 39.

"might answer back, thus upsetting": Serpell, *The Domestic Dog,* pp. 247–48.

"pneumonia, kennel cough, giardia": Andrew J. Manuse, "Rash of Sick Pups Sparks Kennel Investigation," *Boston Herald,* July 22, 2006.

University of Pennsylvania study: D. Duffy et al., "Breed Differences in Canine Aggression," *Applied Animal Behaviour Science* 114 nos. 3–4 (2008): 441–60.

But after a series of well-publicized maulings in 1987: Peter Applebome, "Series of Pit

Bull Attacks Stirs a Clamor for Laws,"
New York Times, July 12, 1987.

studies have linked Pit Bull ownership:
J. Barnes et al., "Ownership of High-Risk
('Vicious') Dogs as a Marker for Deviant
Behaviors," *Journal of Interpersonal Vio-
lence* 21, no. 12 (Dec. 2006): 1616–34.

*"Dobermans, Rottweilers, and German Shep-
herds":* James Hettinger, "Can't Judge a
Bull by the Cover," *Animal Sheltering,*
March/April 2013.

In 1738, a Blue Paul Terrier: Mike Homan,
*The Staffordshire Bull Terrier in History and
Sport* (Devon, UK: Beech Publishing
House, 1988).

"Public Health Implications of Brucella canis
Infections": Prepared by Jim Kazmierczak,
National Association of State Public
Health Veterinarians.

"who had been profiled": Beth Duckett,
"Whisperer Helps with Pets," *Arizona Republic,* Nov. 30, 2009.

"We took refuge in our bedroom": Penelope
Smith, *Animal Talk* (New York: Atria,
2008), p. 20.

through the earth's magnetic fields: Dennis
Moore, " 'Pet Psychic' Unmasks Traumas,
True feelings," *USA Today,* June 3, 2002.

"I'm a good dog": Joan Ranquet, *Communication with All Life: Revelations of an*

Animal Communicator (Carlsbad, NM: Hay House, 2007), p. 47.

"dogs can talk about": Peter Applebome, "Talking to Animals in Their Frequency, and Sniffing," *New York Times,* Jan. 18, 2006.

Chapter 8

In a 2011 article for The Fix: Nic Sheff, "A Dog's Life," *The Fix,* Oct. 15, 2011.

"Walking in front of your dog": Cesar Millan, "6 Tips for Mastering the Dog Walk," *Cesar's Way,* July 2, 2013.

In a 2006 op-ed: Mark Derr, "Pack of Lies," *New York Times,* Aug. 31, 2006.

"hold [the dog] suspended until": William Koehler, *The Koehler Method of Dog Training* (New York: Howell Book House, 1996).

"He'll come with you": Ibid.

"There is nothing that says": "Effective Dog Training — Ian Dunbar," FORA.tv, Jan. 12, 2009.

"We are not like plumbers": Jean Donaldson, "Talk Softly and Carry a Carrot or a Big Stick?" *The Woofer Times,* Sept. 2006.

"It's the urge to dominate": Schaffer, *One Nation Under Dog,* p. 198.

Then there's the Tennessee man: Russell

Goldman, "Elton, the 'Gay' Dog, Spared the Gas Chamber," ABC News, Jan. 13, 2013.

"hair-raising drop-offs": "Road Trip: California's Pacific Coast Highway," Travel.NationalGeographic.com.

sometimes, they actually do: Heather Millar, "The Ups and Downs of Highway 1," *Smithsonian,* June 1999.

"Some dogs, I'm told, like to stick around": Armistead Maupin, "Kiss Patrol," in Michael J. Rosen, ed., *Dog People: Writers and Artists on Canine Companionship* (New York: Artisan, 1995), p. 130.

"Usually they'll hold off": William Yardley, "Oregon Wants 'Dog Friendly' to Be Less So," *New York Times,* Sept. 2, 2009.

the "craziest, most badly behaved sons of bitches": Tara Palmeri, "Liars Use Phony Vests and ID Tags to Get Fake Service Dogs into Posh New York Restaurants," *New York Post,* Aug. 12, 2013.

"These are some of the most loyal": John Weatherford, "Bitchin' News: Experts Say Dogs Owned by Homeless People Don't Have It So Bad," *Willamette Week* (OR), May 24, 2006.

In one study, more than 90 percent: R. Singer et al., "Dilemmas Associated with Rehousing Homeless People Who Have Companion Animals," *Psychological Reports* 77 (1995): 851–57.

"Anyone who has had to sleep": Lars Eighner, *Travels with Lizbeth* (New York: Fawcett Columbine, 1993), p. xiii.

Chapter 9

"the frantic bustle of America": Steinbeck, *Travels with Charley* (New York: Penguin, 2012), p. 115.

"This is a dog of peace and tranquility": Ibid.

"Don't go in there!": Lee H. Whittlesey, *Death in Yellowstone: Accidents and Foolhardiness in the First National Park* (Maryland: Roberts Rinehart, 1995), p. 3.

"Abuse directed against animals": Frank Ascione and Phil Arkow, *Child Abuse, Domestic Violence, and Animal Abuse* (West Lafayette, IN: Purdue University Press, 1999) p. xvi.

Though men are more likely than women: Joni Johnston, "Female Animal Abusers," *Psychology Today,* July 2, 2012.

In an examination of thousands: Kathleen Gerbasi, "Gender and Nonhuman Animal Cruelty Convictions: Data from Pet-Abuse.com," *Society and Animals,* Oct. 2012.

A 2002 task force organized by the Depart-

ment of Education: The Final Report and Findings of the Safe School Initiative: Implications for the Prevention of School Attacks in the United States, May 2012.

"The awkward fact is that most wanton animal cruelty": Herzog, Some We Love, Some We Hate, Some We Eat, p. 34.

Several studies of college students: A. Arluke et al., "The Relationship of Animal Abuse to Violence and Other Forms of Antisocial Behavior," Journal of Interpersonal Violence 14, no. 9 (1999): 963–75.

"I believe simply from enjoying the power": Colin Blakemore, "Darwin Understood the Need for Animal Tests," The Times (London), Feb. 12, 2009.

Dogs at the time could not: Herzog, Some We Love, Some We Hate, Some We Eat, p. 58.

"I will commit to concentration camps": Ibid.

There was the Montana man: Sanjay Talwani, "Man Charged with Getting Dog Drunk," (Helena, MT) Independent Record, March 12, 2012.

the North Carolina man who dragged: "Dog Dragged Behind Moped for Several Miles," Pet-Abuse.com, April 28, 2012.

the New Jersey woman who attempted: Deborah Marko, "Officials Investigate Surgery on Dog at Home," (Vineland, NJ) Daily

Journal, April 12, 2012.

the two California men who lit fireworks: "Small Dog Rigged with M-80 Firework," Pet-Abuse.com, May 23, 2012.

the St. Louis man who photographed himself: Robert Patrick, "Call to St. Louis Home Results in Charges of Child Porn, Sex with Dog," *St. Louis Post-Dispatch,* May 3, 2012.

Kinsey found that 8 percent of men: Marjorie Garber, *Dog Love* (New York: Touchstone, 1997), p. 150.

Dogs were almost removed: "Our Police Dogs Four in Number," *New York Times,* Sept. 22, 1929.

was "viewed skeptically until she detected explosives": Marc Santora and William Rashbaum, "F.B.I. Dog Is Killed in Raid on Hideaway," *New York Times,* March 14, 2013.

This city of 27,000 has the highest: Tim Jones, "East St. Louis Cops Outgunned as Cuts Let Killers Thrive," *Bloomberg,* Jan. 4, 2013.

"genetic castaways on the third rail": Melinda Roth, *The Man Who Talks to Dogs: The Story of Randy Grim and His Fight to Save America's Abandoned Dogs* (New York: Thomas Dunne, 2002), p. 10.

"I've hosted parties I did not attend": Randy

Grim with Melinda Roth, *Don't Dump the Dog: Outrageous Stories and Simple Solutions to Your Worst Dog Behavior Problems* (New York: W. W. Norton, 2009).

"only the brave dare walk its streets": "The East St. Louis Blues," *Time,* April 11, 1969.

The deaths had been widely reported: "Reward Doubled for Dog Torture Suspects," Fox2now.com, May 15, 2012.

"I sing the mangy dog, the pitiful, the homeless dog": Charles Baudelaire, *Paris Spleen* (New York: New Directions, 1970), p. 105.

"of a misfiring of our parental instincts": Herzog, *Some We Love, Some We Hate, Some We Eat,* p. 92.

"From the meme's-eye view": Ibid., p. 93.

"the Japanese began to emulate aspects": Ibid., p. 94.

ABOUT THE AUTHOR

Benoit Denizet-Lewis is a writer with *The New York Times Magazine* and an assistant professor of writing and publishing at Emerson College. He is the author of *America Anonymous: Eight Addicts in Search of a Life*, and has contributed to *Sports Illustrated*, *The New Republic*, *Details*, *Slate*, *Salon*, *Out*, and many others. Denizet-Lewis lives in Jamaica Plain, Massachusetts.

The employees of Thorndike Press hope you have enjoyed this Large Print book. All our Thorndike, Wheeler, and Kennebec Large Print titles are designed for easy reading, and all our books are made to last. Other Thorndike Press Large Print books are available at your library, through selected bookstores, or directly from us.

For information about titles, please call:
  (800) 223-1244

or visit our Web site at:
  http://gale.cengage.com/thorndike

To share your comments, please write:
Publisher
Thorndike Press
10 Water St., Suite 310
Waterville, ME 04901